# GREEN HOMES

## YOUR COMPLETE GUIDE TO CREATING A LOW-ENERGY HOME

28 CASE STUDIES
VITAL COMPONENTS OF AN ECO HOUSE
A TO Z OF GREEN HOMES
GO GREEN QUICK REFERENCE GUIDE

HOMEBUILDING & RENOVATING MAGAZINE

# GREEN HOMES

an ovolo book

**Ovolo Publishing Ltd**
**1 The Granary, Brook Farm,**
**Ellington, Huntingdon,**
**Cambridgeshire**
**PE28 0AE**



ISBN: 978 1 905959 006



All of the material in this book has previously appeared in Homebuilding & Renovating magazine – Britain's best-selling monthly for selfbuilders and renovators (www.homebuilding.co.uk).

Book Design: Gill Lockhart
Publisher: Mark Neeter

This edition first published in the UK by Ovolo Publishing Ltd, 2008
Printed in China

For more information about our books please visit:
www.ovolobooks.co.uk
email: info@ovolobooks.co.uk

To purchase books about building and home interest please vist:
www.buildingbooksdirect.com
email: info@bookdirectlimited.com
or call: 01480 893833 (24 hours)

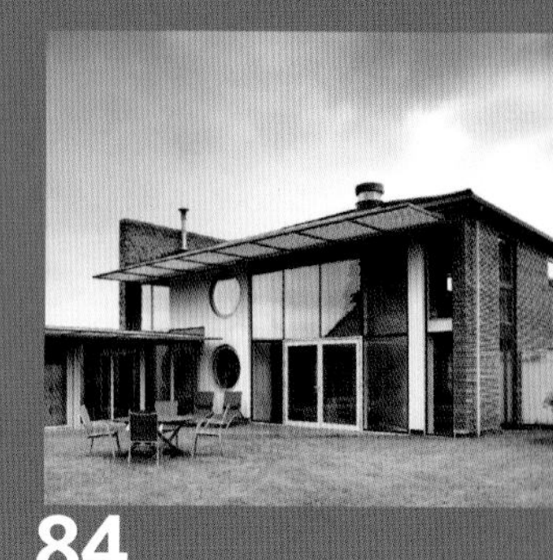


# CONTENTS

56

36

112

122

216

BREAD

# FOREWORD

The Government has established that all new homes built in the UK must achieve Zero Carbon status by 2016 (and by 2011 in Wales). As a result the nation's homebuilders, from the big-name developers through to individual self-builders, are left wondering quite how this will affect the construction of new homes.

Whilst the official Code for Sustainable Homes together with the Building Regulations may help clarify the situation, for many homebuilders the situation may still be less than clear. That's why this unique book – part manual, part companion, part inspiration – has been created to guide you through both the regulatory maze and show just what spectacular buildings can be created.

Building a 'green' home is not just about ticking boxes and deciding which of the increasing number of add-on 'green' features you want – it's about designing and constructing a home that places sustainability at its very core. From employing site-specific solutions such as orientation through to creative design, the use of natural, sustainable materials and, of course, the twin tenets of the green movement: energy efficiency and generation.

The great news for builders of sustainable homes is, as this book shows, that they can not only look fantastic but also cost very little to run. I've seen new Zero Carbon-standard houses that have an annual heating bill of just £30 and enjoy a premium on the resale market thanks to the effects of the Energy Performance Certificate. With higher energy prices and uncertainties over future energy security it now makes sense, from both environmental and affordability arguments, to go green.

Selfbuilders, of course, have been at the forefront of the green homes revolution for years. As with most that's new in housing, selfbuilders are amongst the first to embrace emerging technology and the 28 individual projects in this book do so in brilliantly successfully ways.

The case studies in this book provide an inspiration for us all to go and create new homes that are affordable, desirable and sustainable.

"BUILDING A 'GREEN' HOME IS NOT JUST ABOUT TICKING BOXES, IT'S ABOUT DESIGNING AND CONSTRUCTING A HOME THAT PLACES SUSTAINABILITY AT ITS VERY CORE."

*Jason Orme, Editor, Homebuilding & Renovating magazine*

# VITAL COMPONENTS OF AN ECO HOUSE

Jason Orme introduces the essentials of any self-respecting eco house.

## RAINWATER HARVESTING SYSTEM

Using the rain we get in this country to supplement a home's existing water supply is a relatively straightforward way to reduce your home's impact on the environment. A rain harvesting system costs around £1,500 to install, which includes underground storage tanks, filters and a pump to take the water back to the house. In theory, a large enough tank will supply enough water for a whole house. In reality, harvested rainwater is most commonly used for flushing toilets and watering the garden. The simplest form of rainwater harvesting is the humble water butt to which you can fit pumps to them to allow 100% of your saved rainwater to be hosed onto the garden.

## UNDERFLOOR HEATING

Installing an underfloor heating system is an energy-efficient thing to do. Because the emitter (the floor) has a larger surface area than the standard radiator, it requires the water to be heated to a lower temperature than would otherwise be the case. Installation costs are similar to those for a top-end radiator-based system and running costs are between 10-30% cheaper.

## GREY WATER RECYCLING

Grey water, the waste water from showers, basins and washing machines, can be recycled for non-potable uses such as flushing the toilet and watering the garden. The 'grey' water is treated with cleaning agents and passed through a carbon filter. Less storage is needed than with rainwater harvesting because the supply of water is more regular than rainfall.

## HEAT PUMP

The ground source heat pump works in the same way as a fridge, using electricity to power a system in which water is fed through pipes below the surface (you will need to excavate part of your garden to lay the pipes) at which the temperature remains at a constant 10°c. The warmer water is then 'geared' into a smaller amount of water which raises its temperature to around 30-40°c, enough to power an underfloor heating system. Installation costs tend to be between £8-12,000. Payback times are a point of some controversy in the industry, with claims ranging between 8 and 22 years.

## WIND POWER

Wind power is finding favour among homeowners who want to install a reasonably simple technology that can make a difference to their home's energy consumption. Two basic types exist: the stand-alone system which is used to generate electricity for charging batteries to run small electrical applications; and grid-connected systems, which connect the turbine's output to a home's existing mains supply. In this case the energy produced by the turbine overrides the energy from the grid, with the mains electricity a useful backup. A basic system such as Windsave's WS1000 costs around £1,595 + VAT installed, and taking into account grants available, the manufacturers claim it could pay for itself in as little as five years, with a claimed saving of up to 33% off an annual electricity bill.

However, there is controversy over how much wind is actually converted into energy, and the viability of a system in low-wind areas. Excess electricity can be sold back to the grid, which works through requiring a metering system which will need to be read independently. The meter reading charge is around £50, which means that at least 5,000kWh of exported electricity will be expected before earning sufficient to cover the charge. A 1.5kw turbine will produce an average of 3,942kWh per year, saving 3,390 tonnes of $CO^2$ emission.

## SOLAR POWER

Using the sun to provide energy is split into two areas: solar panels, used for heating water, and PV cells, which are used for creating electricity. A heating system tends to cost around £2,000 installed and can usually provide enough hot water all year round – the problem from a 'value' point of view is that it only costs around £100 a year to provide this anyway. PV cells create a more significant amount of electricity which may allow you to sell some of the energy you've created back to the grid.

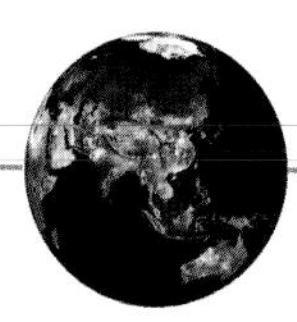

## USE NATURAL BUILDING MATERIALS

Individual housebuilding has a significant impact on the environment, not least of all in the choice of materials that are used in its construction. Materials that require manufacturing inevitably require high levels of energy to produce into workable products, and often give out pollutants as a by-product of the manufacturing process. While plastics are the worst offender, materials such as cement (requiring 1,500kWh/tonne to produce) are very high in embodied energy while steel, copper and aluminium use between 10-30,000kWh/tonne to produce. To give some comparison, a standard house will require some 130,000kWh of embodied energy in its production – about the equivalent of seven years' motoring.

Naturally occurring materials such as lime, sand, stone, slate and timber enjoy very low production costs (of less than 100kWh/tonne) and produce significantly less pollutants in this process, especially if sourced locally.

## INSTALL MORE INSULATION

Modern building regulations specify a relatively high level of insulation in new homes. It is a key ingredient of a green home because it reduces the amount of energy required; good insulation keeps heat in the house (as heat always tries to move from warmer to colder areas). Floors, walls and roofs should have as high levels of insulation specified as possible.

Owners of older properties should be looking to bring the insulation in their homes up to, as close as possible, modern building standards. As up to 33% of the heat produced in your home is lost through walls, cavity wall insulation is something to strongly consider, costing £200-500 and producing savings of around £150 a year on your heating bills.

## ENERGY SAVING LIGHTBULBS

Energy saving lightbulbs use around four times less electricity than a standard lightbulb, so where you would normally use a 40W bulb, you would only need an 8-11W energy saving equivalent. They cost around £3.50 compared to just 50p for a standard bulb, but it should save you around £10 per year on your annual electricity bill.

## A BETTER DESIGN

A successful green house starts with a well thought-out design. The actual house design plays a massive role in deciding how green it is to build and run. For example, all the key living rooms should face south to make the most of the extra light through larger windows – not only will it reduce the need for artificial lighting, but it will also ensure that the passive heat of the sun radiates through the house. Likewise, north-facing elevations should have less fenestration to minimise heat loss through windows to colder spots. Where possible, rooms should have windows on more than one wall – this increases the natural light and is seen to promote wellbeing.

## LOW-E GLASS

Low-emission (low-e) glass has an invisible thin coating of metal oxide, which allows the sun's heat and light to pass through the glass but at the same time blocks heat from leaving the room. This reduces heat loss through windows considerably – by around 30% as opposed to non-coated windows.

## HEAT RECOVERY PASSIVE VENTILATION

A heat recovery system will take the moist, stale air from bathrooms and kitchens (through extract fans) to a heat recovery unit (usually situated in the loft), and pass it through a heat exchanger to recover some 65-70% of the heat usually lost through standard ventilation. They run continuously and are up to 95% efficient, and can easily be fitted to additional heating and cooling systems. This heat is then used to warm up the incoming air into the house, which is then ducted to the living rooms. Payback time is estimated to be around five years, depending on local conditions. A HR unit is very quiet, provides filtered air (perfect for asthma and hayfever sufferers) and does away with the need for trickle vents above windows.

## TIMBER WINDOWS

Timber is the only major building material that is renewable. Using timber for windows (and as much as possible in your house) does not cause global warming; it does not require energy-intensive processing, and it is not problematic to dispose of. Unlike alternative materials it has a low embodied energy and does not require the release of dioxin pollutants and chlorine. Timber uses approximately 300kWh/tonne to produce; uPVC uses approximately 45,000kWh/tonne. On the downside, however, almost all timber used in the UK has to be imported into the country, the transport of which uses up energy. You should always ensure the timber you use comes from sustainable sources – meaning that for every tree felled, another is planted. Preferably, your supplier should be a member of one of the leading sustainable timber campaigns in the UK: Wood for Good or Forests Forever, which attempts to ensure that local people benefit from the income generated by their forests.

## LOW FLUSH WCS

The average household uses 180 litres of water a day to flush toilets, accounting for the largest single use of water in the home (around a third of

'WASTE WATER FROM SHOWERS, BASINS AND WASHING MACHINES CAN BE RECYCLED FOR NON-POTABLE USES SUCH AS FLUSHING THE TOILET AND WATERING THE GARDEN'

TIMBER IS THE ONLY MAJOR BUILDING MATERIAL THAT IS RENEWABLE. USING TIMBER FOR WINDOWS (AND AS MUCH AS POSSIBLE IN YOUR HOUSE) DOES NOT CAUSE GLOBAL WARMING

all the water we use is used this way). Dual flush toilets – providing smaller flushes with different buttons – are an essential starting point for any home, but water-saving or low-flush WCs are an even better way to save water. They reduce the flushes by around half which, thanks to a specific bowl and drop valve design, is still more than effective. Some new products can be fitted retrospectively to existing toilets.

## NON-TOXIC PAINTS

Conventional paints contain a range of chemicals (VOCs) that contain harmful pollutants, require energy to produce (1 litre of paint can result in up to 30 units of toxic waste) but can cause health problems. Regular painters can suffer headaches, skin problems and even find their central nervous system attacked (painters' dementia is a recognised disease in Denmark). Safe eco paints, however, use natural ingredients such as vegetable oils, clays and plant extracts to produce a durable covering with minimal waste.

## NATURAL LIGHTING

Houses should be designed to allow in as much natural light as possible. However, in certain parts of the house, on certain sites, this can be difficult to achieve and the conventional solution is to rely on artificial lighting. Light pipes are the green solution – a small dome situated on the roof ('conservation' models replace the dome with a flat panel) gathers light and channels it down through mirrored reflective tube ducting into the room of your choice. It is perfect for downstairs bathrooms, corridors or basements. The light is free and works just as well in overcast conditions. It is so efficient that the companies pioneering this technology are dealing with ways to ensure the light can be blocked out at night. Prices start from around £180.

## BUY LOCAL BUILDING MATERIALS

Where possible, you should choose building materials that have been sourced and manufactured locally. It's good from a green perspective because it helps to support local communities but it also cuts down on the amount of energy used in transporting the materials to your site. Locally-sourced materials are much more likely to be in keeping with the overall style of house that you would want to build. Where possible, avoid importing items from abroad, as the shipping involved uses a lot of energy and it's difficult to be assured of the environmental credentials of the carrier.

## CHOOSE YOUR BOILER WISELY

Boiler efficiencies have been transformed in recent years to reduce the level of $CO^2$ emissions. Even as recently as 2002, there were no minimum standards for boiler efficiency, which tended to hover around the 70% (of fuel turned into hot water) mark; nowadays, boilers can only be used if they are A or B rated on the Government's SEDBUK rating scheme, which means that all boilers in the UK have to be at least 86% efficient – a rating that only condensing boilers currently exceed. Using one will save you up to around £60 a year on your heating bills. One extra issue to take into account is modulation: it is capable of altering the amount of fuel burned off by the boiler. A modulator adjusts the fuel input in order to increase efficiency even further, instead of simply being on or off.

## CHP UNITS

CHP – Combined Heat and Power – units work in a simple, effective and green way. They use gas or oil to drive a generator that produces electricity. The heat from the engine block, oil cooler or exhaust is absorbed by coolant water through a high-efficiency heat exchanger to produce hot water. In rough terms, the production of 1kW of power also produces 2kW of useable heat energy. In many ways, it acts like a normal gas boiler, but saves around 1.5tonnes of CO2 emissions each year, and around £150.

## CHOOSE A GREEN ENERGY TARIFF

In the UK, around two-thirds of electricity is generated by burning coal and gas in power stations. Nuclear power accounts for around an additional quarter, with the rest coming from sources which have no or limited impact on the environment. You can choose a supplier (or a tariff from your regular supplier) which either ensures that the electricity coming to your home is from a green source or that the money you spend on energy with the supplier is being channelled into green projects (each supplier offers different 'green' factors with its tariffs). The best known, and most mainstream of these tariffs is the 'Juice' tariff from nPower (developed in collaboration with Greenpeace) – you pay exactly the same price for your electricity but you get it from the North Hoyle offshore windfarm. It's simple and easy to switch to a green tariff and in most cases won't cost you a penny.

## CHOOSE THE RIGHT APPLIANCES

Your appliances – fridge, dishwasher, washing machine etc. – take up a surprisingly large amount of your daily energy, costing between £25-40 a year to run. When choosing appliances you should look at the EU energy label that all appliances have to show — choose A (now up to A++ for refrigeration) rated products that consume less energy and you could save up to 25% of energy. Look out too for appliances that are recommended for energy efficiency by the Energy Saving Trust.

**Confused by all the talk about green homes? We explain the key terms in the ultimate reference guide for self-builders and renovators.**

BY JASON ORME AND MARK BRINKLEY

## AIRTIGHTNESS

'Build it tight, ventilate it right' is one of the new principles of green homebuilding – based on the idea that air loss from a home is an inefficient use of energy. Under Part LIA of the Building Regulations, all new homes must now have an air-leakage test to assess how much air is being lost from the house through openings in the structure. The Regulations require homes to be built with a maximum of 10 air changes an hour, with a target of five (for comparison, homes built before 1960 could have up to 20 air changes an hour). This is particularly an issue at 'weak' areas such as the joining of the roof to the wall structure. Airtightness can be achieved through better building practice, especially in bonding and joints, around sockets and weak points; if achieved successfully, it can reduce fuel costs by up to 80%. Owners of airtight homes report lesser instances of allergy problems.

According to Michael Benfield of self-build and timber frame specialist Benfield ATT: "Greater airtightness is achieved through careful design and scrupulous attention to on-site trade detail. While it is possible to reach present requirements using all masonry construction, it should be noted that, for example, block-makers have had to devise a 'thin joint' system to make this more reachable. Since panelised systems are generally more impermeable than masonry, it follows that it is generally easier for them to attain lower rates of air change. However, even these are subject to possible on-site failures. They need to be properly sealed and caulked to eliminate leakage at all weak points and, like all other systems, can be weakened by careless puncturing of the fabric, eg by follow-on trades inserting electrical wiring, plumbing and waste pipes." Owing to the lack of air movement, mechanical ventilation systems, which automatically extract stale air and replace it with fresh air, must be used when building an airtight home. When combined with a heat-recovery system, which uses the moist warm air extracted from bathrooms (for instance) to produce warm air, the system can also reduce energy consumption.

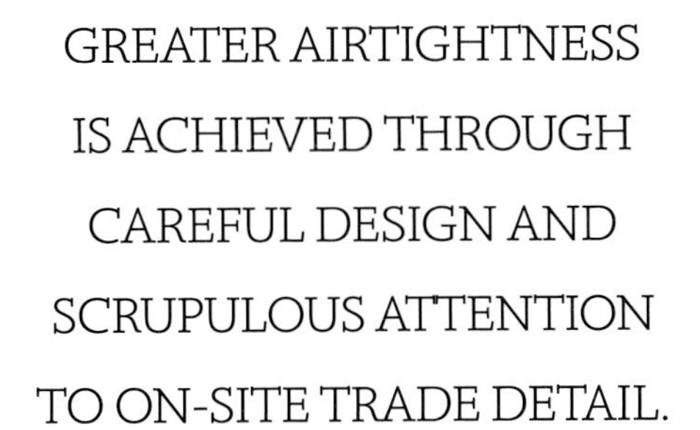

GREATER AIRTIGHTNESS IS ACHIEVED THROUGH CAREFUL DESIGN AND SCRUPULOUS ATTENTION TO ON-SITE TRADE DETAIL.

Benfield ATT Timber Frames: 01291 437050
Super-E (standard for airtight housing): www.super-e.com
Mechanical ventilation systems – Starkey Systems: 01905 611041
www.centralvacuums.co.uk

## AIR-SOURCE HEAT PUMPS

The basic idea behind heat pumps is that they take a small amount of heat from a large source and convert it into a much larger amount of heat in a much smaller medium, typically hot water in a cylinder.

Heat pumps are neither a renewable technology nor an example of microgeneration and are eligible for grant aid. They do promise a much greater heat output than energy input: the key to understanding this is in the ratio known as the Coefficient of Performance (CoP), which indicates how much additional power is being produced for any given power input at a range of temperatures. Most heat pumps boast a CoP of 3.0 or higher – this means that they are producing three times as much energy as they use in operation.

Air-source heat pumps (ASHP) differ from the better-known ground-source heat pumps in that they are much cheaper and simpler to install as they don't require extensive underground pipework to collect heat. Essentially, you get a box not unlike an air conditioning unit, which can be mounted on or near the house, and this draws heat out of the air. The efficiency of the unit varies with the air temperature and on the coldest days an ASHP may struggle to cope with the demand from a house for space heating and hot water. They also tend to be a little noisy: you need to check the decibel levels and think carefully about where it might be best sited. Ground-source heat pumps are quieter in operation but require a large area of garden land to be dug up.

Trianco ActiveAir: 0114 257 2300 www.trianco.co.uk/activair.cfm Nu-Heat: 0800 731 1976 www.nu-heat.co.uk

## BIOMASS

Biomass, or bioenergy, refers to the use of non-fossil fuels to power specialised boilers. Although biomass fuel releases $CO_2$, it's part of the normal carbon cycle and is not adding to atmospheric $CO_2$ – provided your fuel is from a sustainable source, where replanting is taking place.

However, there are often high transport costs involved in getting biomass fuels to the point of consumption, so it suits some sites much better than others. Biomass boilers tend to be aimed at rural houses, which have access to local timber and have sufficient storage space.

In order to address some of these concerns, wood pellet fuels have been developed which are much more readily transportable. They are widely used in Sweden, Denmark and Germany and we now have pellet plants starting up in the UK. Wood pellets are made from timber which is compressed to such a density that it holds the same calorific value as heating oil; better still, it tends to sell for the same price. Pellet fuels are, therefore, being created to trade as an alternative to heating oil.

Biomass and pellet boilers are expensive when compared to fossil-fuelled oil or gas boilers, though the best ones offer levels of automation, which take most of the drudgery away from solid-fuel devices. As with many other renewable technologies, grants are available.

More information is available from the Log Pile Website at www.nef.org.uk/logpile ➤

## COB

An ancient building technique using a mixture of earth, straw, sand and other materials. It is similar to adobe and has been used in the UK for many centuries, particularly in the west. It is exceptionally cheap and while earth is not a great insulator, building with cob does mean creating a very thick solid thermal mass, and it does enjoy high absorption levels of solar energy. It also achieves an exceptionally low carbon footprint.

**Build Something Beautiful: 01404 814270 www.buildsomethingbeautiful.com**
**Devon Earth Building Association: www.devonearthbuilding.com**
**Earthed: www.earthedworld.co.uk**

## CODE FOR SUSTAINABLE HOMES

Introduced in December 2006, this is the key plank of the Government's policy to move all housing to zero carbon by 2016. It is split into six levels, each more demanding. Level 1 is slightly better than the current Building Regulations standards. Code Level 4 is equivalent to the German PassivHaus standard, an ultra-low-energy house that doesn't require any on-site power generation. Code Levels 5 and 6 are variously described as net zero carbon and zero carbon. The Code also has requirements for water-efficiency measures and management of waste and resources.

The Government has indicated a timetable for the adoption of the various Code levels into the Building Regulations, specifically Part L in England. Wales, Scotland and N. Ireland are adopting their own timetables but anticipate them being broadly similar, though Wales has broken ranks by insisting that all new buildings there should be zero carbon by 2011.

| CSH Level | % better than '06 Part L | Comments | Becomes Part L |
|---|---|---|---|
| 1 | 10 | EST Good Practice | |
| 2 | 18 | | |
| 3 | 25 | EST Best Practice | 2010 |
| 4 | 44 | Near PassivHaus | 2013 |
| 5 | 100 | Zero heating and lighting | |
| 6 | 100 | Zero extends to cooking and appliances | 2016 |

**See also Zero-Carbon Homes**

## COMBINED HEAT POWER (CHP)

Imagine a boiler that produces electricity as well as hot water. That, in essence, is what combined heat and power (CHP) plants do. Instead of energy going to waste, it converts most of it into usable power, in the form of electricity. It's been used successfully for many years on large mixed developments where there are shops, offices and homes, but only recently have we seen the advent of single-house solutions, known as micro-CHP where it is designed to replace gas boilers.

However, at the time of writing, no one is supplying micro-CHP into the UK market. The leader in this field is Powergen, who has a micro-CHP unit called WhisperGen, which has been installed on a trial basis in a number of homes. But to date they have all been hand-built in New Zealand and until the company can find a mass production facility closer to home, it is not taking any new orders until 2009.

Looking a little further into the future, we can anticipate the development of fuel cell-based CHP plants which promise to be even more energy efficient than the gas-powered versions of today. At the moment, domestic fuel cell systems are still some way from commercial reality and it may be ten years or more before we see any hydrogen-based fuel cells in the home. CHP as a technology is still evolving, but it seems it is more likely to be suitable for powering district heating schemes rather than individual households.

**More information on WhisperGen is available through Powergen: 0800 096 1160 www.powergen.co.uk**

LOOKING INTO THE FUTURE, WE CAN ANTICIPATE THE DEVELOPMENT OF FUEL CELL-BASED CHP PLANTS WHICH PROMISE TO BE EVEN MORE ENERGY EFFICIENT

## CONDENSING BOILERS

Condensing boilers have extra heat exchangers that extract a greater amount of useable energy from the fuel. The heat exchanger uses the heat in the exhaust gases from the boiler to preheat the water as it enters the boiler. As a result they make use of in excess of 90% of the input fuel – as opposed to just 60-80% in standard boilers (and much less in older models). They can be either combi or standard system boilers, and come in gas or oil models.

Under changes to the Building Regulations in 2004 and 2007, it is now compulsory for all new boilers (in new builds or renovations) to achieve an 'A' or 'B' rating on the Sedbuk scale of boiler efficiency — which means that boilers must now achieve a minimum 85% efficiency rating. In practice, this means that all new boilers must now be condensing.

**Sedbuk boiler efficiency database: www.sedbuk.com**

## DUAL-FLUSH LOOS

Water Supply Regulations state that a toilet can have a maximum flush of six litres (it used to be 13 litres in the 1960s), but self-builders and renovators can save themselves water (and money, as WCs account for up to 40% of domestic water use) through a dual-flush model – WCs that have two different flush volumes (usually six litres and four litres). In

practice, the minimum flush volume is two litres, which is quite common in Scandinavia, although most UK experts believe the practical minimum to be three. It is also possible to retrofit devices in existing cisterns to reduce flush volumes.

Hippo: www.hippo-the-watersaver.co.uk
Interflush: 0845 045 0276 www.interflush.co.uk

## DRIVES

Your driveway – and much of the hard landscaping outside your house – potentially has a large environmental impact, both in its production and in the impact it has on storm sewers in excessive weather conditions. Current practices which prevent rainwater from entering the subsoil (reducing groundwater recharge) mean that infiltration decreases, base flows in streams are decreased and some small streams may dry up – which has significant effects on water supply for everyone. You can reduce it by investing in something like a GrassGrid driveway, which is an interlocking mesh system (similar, in fact, to the look of a dry ski slope) which combines the hard-wearing properties of concrete but allows grass to grow in between – meaning less run-off into the drainage system, a more attractive finish, and a smaller carbon footprint than concrete or asphalt.

Alternatively, it is possible to pick up 'hard' materials that allow air and water to pass through. So-called 'porous asphalt' and 'porous concrete' are at early stages of development but have been used in commercial applications.

Grasscrete: 01924 379443
GrassGrid driveway – Charcon: 01335 372222.
Perfo UK: 01992 522797 Source Control: 01283 509021

## ENERGY-EFFICIENT LIGHTING

Traditional light bulbs waste a lot of their energy by turning it into heat rather than light, and while dimmer switches can reduce the feed to light fittings, there is still some waste..

Compact fluorescent lamps (CFLs) are a form of energy-saving light bulb which look like small tubes. They do, however, contain small amounts of mercury and users occasionally complain of poor light quality. Energy-saving CFL light bulbs use a quarter of the electricity of ordinary bulbs to generate the same amount of light – so where you would normally use a 60W bulb, you will only need a 13-18W energy-saving equivalent. In addition, they last up to 12 times longer than ordinary bulbs and can save £9 per year in electricity. The good news is that, thanks to advances in technology, many of the best bulbs now enjoy 'instant start-ups' and a better light quality.

Halogen lights run on a lower voltage and require a transformer. They also tend to blow quite often so manufacturers recommend not turning halogen lights on/off as often as you might with other bulbs.

An LED light only consumes 3W of energy yet has the same light output as a 20W bulb. They use less than a quarter of the electricity that fluorescent lighting does. They also have incredibly long lifespans – 100 times more than fluorescents – in fact, they mostly fail through gradual dimming rather than burn out. They also require transformers.

Osram: www.osram.co.uk, IKEA: www.ikea.com
Philips: www.philips.com, GE Lighting: www.gelighting.com

## FORMWORK

Insulating concrete formwork (ICF) is a building system that provides formwork for in-situ concrete structures. The formwork is left in place permanently and acts as thermal insulation. ICF consists of twin-walled expanded polystyrene (EPS) blocks that are built up to form walls which are then filled with ready-mixed concrete. The combination of EPS insulation and in-situ concrete produces a wall U-value of under 0.2W/m²K. The concrete has a high thermal mass which means that it absorbs and stores heat or coolness much better than regular walls.

ICF is an emerging form of housebuilding technology thanks to its ease of use and highly insulative qualities.

Insulating Concrete Formwork Association: 0700 450 0500 www.icfinfo.org.uk StyroStone International: 0871 789 7678 www.styrostone.com
Beco Wallform: 01724 747576 www.becowallform.co.uk

BOILERS MUST NOW ACHIEVE A MINIMUM 85% EFFICIENCY RATING. THIS MEANS THAT ALL NEW BOILERS MUST NOW BE CONDENSING

## GLAZING

Windows are weak elements when it comes to keeping heat in the building. Glass is a highly conductive material. The thermal performance of windows can be improved by increasing the number of glazing layers, increasing the cavity between the layers, adding a coating to the glazing, and filling the cavity itself with gasses. In addition, the frame's performance can be enhanced by using materials with lower thermal conductivity (e.g. wood) and creating thermal breaks in the frame.

The U-values for single, double and triple glazing are 4.8, 2.8 and 2.1W/m²K respectively (with a 12mm air gap in between).
The U-values for double glazed units with 6, 12 and 16mm air gaps are 3.1, 2.8 and 2.7W/m²K respectively.

By coating glass during manufacture with a fine layer of metal oxide, thermal radiation can be better controlled through window glazing. As ➤

emissivity coatings can either be 'hard' or 'soft', the warmth of the sunlight can be let through without letting it escape from the house.

Reductions in the U-value of glazed units can be made by replacing the dry air with argon, which has lower thermal conductivity than air.

Pilkington: 01745 536500 www.pilkington.com

## GRANTS

Householders can get Government assistance to reduce the high capital cost of energy-generating features. In the UK, grants are administered by the Low Carbon Buildings Programme. Heat pumps, for example, get a maximum grant of £1,200. The system has been mired in controversy for many months owing to massive take-up of the monthly allowance (funds usually ran out on the first day of each month) and, after temporarily closing the scheme down, the DTI recently relaunched the scheme, increasing the amount of money on offer to £18m.

However, in order to spread the available pot further, the maximum grant offered to householders has been reduced from £15,000 to £2,500 (per household, not per feature). "This will particularly hit PV and wind, which have been two of the most popular elements of the Programme," says a spokesman for the Renewable Energy Association. "Demand has collapsed following the DTI's decision to suspend the Programme in March, and with these unnecessary funding cuts it is unlikely to recover." The average PV system grant, for example, has been about £6,300, so the new rules will reduce this by 60%.

In Scotland, the situation is more generous. Funding for householders is set at 30% of the installed cost of the feature up to £4,000, and householders can also apply for separate grants for two different technologies.

Northern Irish readers benefit from an £8m funded grant system that allows householders to apply for up to two different installations on one building, with individual grants varying by feature with a limit of £15,000 on PV panels and £3,000 on heat pumps. Providing they are not applying for the same technology, Northern Ireland residents can apply for additional grants through the Low Carbon Buildings Programme.

Irish citizens can enjoy grants up to e3,600 for solar thermal panels and e6,500 for a ground-source heat pump, and can apply for two grants where technologies complement each other.

The difference to the end user of these grant variances can be significant. A person buying a ground-source heat pump costing £9,000 and £2,000 worth of solar thermal panels would end up spending a total of £9,400 in England and Wales, £7,700 in Scotland and just £7,000 in Northern Ireland, where the grant situation for solar thermal panels is particularly advantageous, allowing £1,125 regardless of size up to an overall 50% limit.

England and Wales: 0800 915 0990 www.lowcarbonbuildings.org.uk
Scotland: www.energysavingtrust.org.uk/schri Northern Ireland: 0800 023 4077 www.actionrenewables.org Ireland: www.sei.ie

WINDOWS CAN BE IMPROVED BY INCREASING THE NUMBER OF GLAZING LAYERS, THE CAVITY BETWEEN THE LAYERS, ADDING A COATING TO THE GLAZING, AND FILLING THE CAVITY ITSELF WITH GASSES

## GREEN ROOFS

A green roof is a roof that is covered with vegetation planted over a waterproofing membrane, filter sheet, root membrane and timber frame. They are seen as energy efficient because they add thermal mass to a roof structure and reduce storm-water run-off, although the insulative qualities of soil are not seen as terribly impressive. The vegetation is usually sedum, a form of leaf. Green roofs tend to work best with pitched structures, as the lack of a water run-off can cause issues of loading and potential failure in flat roofs. Pitches should be no higher than 15 degrees as anything steeper will cause soil run-off.

Living Roofs: www.livingroofs.org Safeguard Europe (membranes ideal for green roofs): 01403 210204 www.safeguardeurope.com

## GREY WATER RECYCLING

Two types of water can be recycled – rainwater and grey water. Grey water recycling systems work by taking water draining from the bathtub, washbasins and showers, filtering out the suspensions and storing it in a small holding tank. The water is then pumped into a header tank in the loft where it is treated with disinfectant and used to provide water in the WC cisterns and can cut water bills by up to 40%. Units cost around £1,000. A storage tank of one cubic metre (1,000 litres) is considered adequate for households.

Aquaco: 01892 506851 www.aquaco.co.uk CPM Group: 0117 981 2791 www.cpm-group.com Freewater UK: 01522 720862 www.freewateruk.co.uk Biotank: 01277 889333 www.biotank.co.uk

## GROUND-SOURCE HEAT PUMPS

The alternative technology best adapted to underfloor heating is the ground-source heat pump (GSHP), because it works most efficiently at low temperatures. It is not surprising, therefore, that self-builders, who have been installing underfloor heating for many years, should now be

turning to GSHP in large numbers as well.

The ground-source heat pump draws low-grade heat from large volumes of liquid being pumped around the garden and converts this into smaller volumes of higher-grade heat that is available for space heating and hot water. GSHP has been around in the UK since the 1950s, but the availability of cheap oil and gas in the UK held back its roll-out. However, the oil price hikes of the past three years have changed all that and now self-builders are installing GSHPs in increasing numbers, confident there will be a sensible economic payback. There needs to be. Ground-source heat pumps are relatively expensive to install – typically, costs are between £8,000 and £10,000 – but promise very low running costs because they produce three or four units of heat from every unit used to operate the system. However, the fuel that they operate on is electricity and the amount of electricity needed to power up a GSHP plant is much greater than can be delivered by photovoltaics or wind turbines.

Some argue that GSHPs really aren't that green. But, as long as the Coefficient of Performance (the amount of power out over the amount of power in) is greater than 2.0, and a well-designed system should achieve far better than this, a GSHP will produce less $CO_2$ than a condensing boiler. GSHPs are eligible for grants under the Low Carbon Buildings Programme.

THE GROUND-SOURCE HEAT PUMP DRAWS LOW-GRADE HEAT FROM LARGE VOLUMES OF LIQUID BEING PUMPED AROUND THE GARDEN AND CONVERTS THIS INTO SMALLER VOLUMES OF HIGHER-GRADE HEAT

Worcester Bosch: 01905 754624
www.worcester-bosch.co.uk
ECO Heat Pumps: 0114 296 2227
www.ecoheatpumps.co.uk
Hidden Energy: 01327 872435
www.hiddenenergy.co.uk Ice Energy: 01865 882202 www.iceenergy.co.uk
Thormec: 0845 052 5325 www.thormec.co.uk
Viessmann: 01952 675060 www.viessmann.co.uk

## HEALTHY HOMES

There is a school of thought which suggests that the buildings we inhabit can have a significant impact on our overall wellbeing. So-called 'Sick Building Syndrome' can manifest itself in a range of symptoms from headaches and fatigue to skin irritations and coughs, and it is important to consider the effect that your home can have on your general wellbeing. A home can make you ill in obvious ways – from the deleterious effects of asbestos through to illnesses associated with leaky flues – but it can also be a lot more insidious. Many building materials have high levels of volatile organic compounds (VOCs) — high concentrations of which can cause cancer. Exponents of 'Healthy Home' philosophy advocate removing toxins and chemicals from any of the building materials entering your home – toxic paints, certain forms of insulation, and so on – as well as paying attention to the potentially harmful effects of radiation from electrical implements.

## HEMP

In the 18th century, hemp was used in everything from clothes to paper – and contributed significantly as a structural material. The technique is undergoing a revival in interest due to its insulative properties: when mixed with lime, it can provide a U-value of around 0.45 or lower when used in 230mm-thick solid walls. It grows to maturity in four months, requires no pesticides, puts nutrients back into the soil and has a high yield. It is highly breathable and has excellent thermal qualities.

## HYDRO

Harnessing the power of fast-flowing nearby rivers and streams is, on paper at least, one of the most effective and reliable ways of generating energy using renewable methods. It's very efficient in the way it converts available energy into electricity – up to 90%, far and away better than its alternatives – and is easy to predict, in a way that solar and wind is not; it's also constant and doesn't tend to change in its output, although there is a slight fall away in summer. The key elements in a successful hydro plant are flow and head: you obviously need a fast flow but you also need, to coin a phrase, good head — the vertical fall of the water from upstream to downstream. For this reason, many of the most successful one-off domestic installations are based around conversions of former mills into new homes.

Information on flow measurements from the Environment Agency.
For a comprehensive guide to small-scale domestic hydropower, visit the British Hydropower Association at www.british-hydro.org/mini-hydro

## INSULATION

While the Building Regulations provide minimum requirements for insulation for external walls, floors, ceilings and roofs, exceeding these standards is a great way to make your home as energy efficient as possible. Current regulations mean that at least 100mm of standard wool-type insulation is required, for instance, on an external wall (this is reduced to 70mm for the polyurethane-type boards); however, super-efficient homes are now being built, particularly in Europe, with a cavity of up to 300mm being filled with insulation (the extra-wide cavity does require special wall ties and is not really yet designed for UK construction standards). It goes without saying that, especially considering the minimal upfront capital cost implications of extra insulation, it pays to specify as much as possible. In addition, retrofitting insulation in older houses makes common sense. An estimated initial outlay of around £300, from fitting cavity wall insulation, for instance, results in annual savings on your heating bill of around a third (£130-£160 a year).

While the traditional type of insulation – glass or rockwool – is perfectly adequate in horizontal situations, and is industry standard for external walls, there is an argument to suggest that rigid boards – produced by the likes of Kingspan – are more likely to stand the test of time. In addition, it is possible to specify construction systems with insulation already built into the structure – through structural insulated panels (SIPs) or insulated concrete formwork (ICF).

Self-builders and renovators particularly interested in natural alternatives (standard forms of insulation have never had a particularly great record (asbestos) when it comes to being brilliantly healthy) can now opt for a range of insulation products that don't use synthetic materials. Sheep's wool insulation is probably best known and while it is more expensive than its synthetic wool counterparts, it is a lot easier to work with – glass fibre and mineral wool requires careful handling and face masks to reduce skin irritation during insulation. Cellulose – the most common renewable insulation material available in the UK – is made from post and pre-consumer waste paper and is available in loose-fill bags or boards of varying thickness. Wood fibres and hemp are other alternatives.

CELLULOSE, THE MOST COMMON RENEWABLE INSULATION MATERIAL AVAILABLE IN THE UK, IS MADE FROM POST AND PRE-CONSUMER WASTE PAPER

**Glass/rockwool insulation – Rockwool: 01656 862621**
**Rigid urethane boards – Celotex: 01473 822093; Kingspan: 0870 850 8555**
**Sheep's wool insulation – Thermafleece, available from Second Nature: 01768 486285 www.secondnatureuk.com Homatherm and Thermo-Hemp available from Ecological Building Systems: www.ecologicalbuildingsystems.com**

## LOW-E

Low emissivity (low-E) refers to a microscopic, virtually invisible metallic oxide coat applied to the inner face of a double-glazed window unit and reduces heat loss through the glass. As most of the heat transfer in multi-layer glazing is thermal radiation from a warm pane of glass to a cold pane, coating a pane of glass with a low-E material and facing that coating into the gap between the glass layers blocks a significant amount of this radiant heat transfer, thus making the window unit more efficient, and reducing heat loss.

See also Glazing

## LOW-FLOW SHOWERS

It's common wisdom that showers use much less water than baths – but a power shower can use more water than a bath (it depends, of course, on how long you are in there). Many showers are now sold with flow restrictors or low-flow settings, which can significantly reduce water usage within a household.

It's also possible to retrofit low-flow shower heads – the aerating models mix air into the water stream and tend to give a more 'full-flow' experience. Flows can get down to around four litres a minute, from around 20 litres a minute for top power showers, massively reducing a home's water usage.

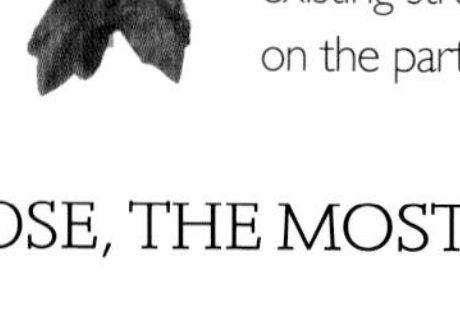

## MICROGENERATION

The generation of power – heat, water or electricity – by homeowners, as opposed to communities and governments (the existing structure). There is a huge push to encourage microgeneration on the part of individual householders and the Government believes that up to around 50% of a home's electricity needs could be met through a variety of microgeneration technologies.

To this end it is currently reviewing planning policies to take the installation of such technologies – solar panels, wind turbines and so on – out of the planning process altogether.

On the other side of the debate, several industry experts believe that renewable energies enjoy much greater efficiencies on a macro level, preferring to see large-scale wind farms and solar farms instead.

## MORTGAGES

Almost every other aspect of the housebuilding and home improvement sector has a green angle – and finance is no different. If you're intending to build a house with several low-energy features, you should consider contacting the Ecology Building Society, a lender that aims to promote sustainability through its lending strategy, and which uses the environmental impact of the project as a starting point. The Ecology is particularly keen to lend on projects that restore or even save derelict properties, homes or otherwise, and is likely to lend on projects that conventional societies will not consider. In addition, you could choose to take out one of the growing number of eco mortgages – for example, the Co-operative's Eco Mortgage offsets 20% of every customer's home's $CO_2$ emissions each year. Green mortgages – where preferential lending rates are

linked to a home's eco credentials – are quite popular in Europe, and the Government is currently considering introducing incentives such as Council Tax rebates for energy-efficient properties.

Ecology Building Society: 0845 674 5566 www.ecologybuildingsociety.co.uk
Co-op Bank: 0845 721 2212 www.co-operativebank.co.uk

## NATURAL PAINTS AND OILS

Headaches and allergies are common side effects of decorating with ordinary paints – which is why, at the very least, you should always keep a room you are decorating well ventilated. In addition to the damaging fumes given off by ordinary paints, the petrochemicals involved in their manufacture are carbon intensive and can pollute the existing water systems. The effects of toxins on regular painters are at an early stage of investigation, but are well recognised – in Denmark, for instance, 'Painter's Dementia' is a recognised disease. As a result, in January 2006, new regulations came into existence setting maximum VOC levels for all paint manufacturers. The toxins in the paint can also off-gas for some time afterwards, affecting you not just when you paint, but when you use the room.

For this reason, an increasingly critical feature of an eco home is the use of natural paints, which contain no solvents, toxins or VOCs. They're also largely odour free, so much more pleasant to use.

Interestingly, a new product recently developed in this market is Insulating Matt Wallpaint, by Ecos. It claims to be able to offer between 10-30% efficiencies as opposed to standard paints and can, of course, help to keep rooms cool in summer as well as warm in winter.

Ecos Organic Paints: www.ecosorganicpaints.com
earthBorn paints: 01928 734171 www.earthbornpaints.co.uk
Auro: 01452 772020 www.auro.co.uk
Nutshell Natural Paints: 01392 421535
Osmo (Oils): 01296 481220 www.osmouk.com

## ORIENTATION

Through appropriate orientation – the positioning of a house on its site – it is possible to maximise the beneficial effects of solar gain. Specifically designed passive solar housing schemes have demonstrated annual energy savings in excess of 10%, although 3-5% is more common.

The general principle is to locate significant amounts of glazing on south-facing elevations, therefore maximising the capture of solar radiation, while minimising its loss through colder north-facing walls by having fewer, smaller windows. There is even a new house currently going through planning permission in Derbyshire that is being built on rollers, so that it can turn to face the sun at all times.

THE GOVERNMENT BELIEVES THAT UP TO AROUND 50% OF A HOME'S ELECTRICITY NEEDS COULD BE MET THROUGH A VARIETY OF MICROGENERATION TECHNOLOGIES

## PV (OR PHOTOVOLTAIC) MODULES

Photovoltaic or PV roof arrays generate electricity from sunlight. The original technology was first developed to help power satellites in space, but PV panels started appearing experimentally on roofs in the early 1990s. Since then, they have grown rapidly in popularity and prices have become more accessible.

Germany in particular has pioneered the introduction of PV arrays by offering large subsidies to homeowners who sell home-generated electricity back into the grid system. In other countries the take-up has been much slower, due entirely to the large installation costs: in Britain, the installation costs are around £1,000 per square metre, enough to produce a peak output of 120W of electricity, or around 100kWh per annum. Whilst there are grants available to help reduce the installation costs, the Government here is in the habit of continually tinkering with them, which does nothing to create a stable market or to inspire confidence in would-be installers.

Nevertheless, many commentators predict that PV arrays will become steadily cheaper over the coming years as global production ramps up, driven partly by policies such as our Code for Sustainable Homes, which requires on-site electricity production on all new homes. Photovoltaics may be expensive but they remain the easiest way to generate renewable electricity on site.

The technology itself is also evolving. We now have individual PV slates which can be built into roofs without the need for a surface-mounted panel. And we are also seeing the introduction of photovoltaic glass, which may in time come to replace large glazed screen areas. The electricity produced is DC and it needs to be converted into AC for use around the house or export to the grid. This is done via an inverter, which is usually mounted in the utility room or garage. Export to the grid in the UK is becoming easier, though the prices paid for homemade electricity are still usually lower than those paid for power bought in.

Solar Century: 020 7803 0100 www.solarcentury.com
PV Systems: 029 2082 0910 www.pvsystems.com
Viessmann: 01952 675060 www.viessmann.co.uk
For more information visit the British Photovoltaic Association at www.pv-uk.org.uk

## PASSIVHAUS

The term 'PassivHaus' refers to a specific construction standard for residential buildings which have excellent comfort conditions. PassivHaus dwellings – the model is used as a basis for the Government's definition of a zero-carbon home – use a range of low-energy features and principles to achieve ultra-low carbon emissions. These include excellent levels of insulation and minimal thermal bridges; utilisation of solar gain; high levels of airtightness; and a whole-house mechanical ventilation system with heat recovery. As a result of these features, minimal additional heating is required – something in the region of 15kWh/m$^2$ per year (compared to around four times this amount in conventional housing). The total primary energy use in a PassivHaus is less than 120kWh/m$^2$ per year.

Over 6,000 PassivHaus dwellings have been built across Europe and some are being undertaken in the UK.

For more information visit www.passivhaus.org.uk

## RECLAMATION

The reuse of building materials is the greenest possible way of constructing or renovating a home. It is possible to buy second-hand bricks, roof tiles, joinery, insulation, plasterboards and pretty much everything else you can think of. Ensure that the materials meet modern building standards, however, and check to see how they will fit in with the rest of your project – for instance, reclaimed radiators may well need some form of adaptation to join up to modern heating systems. Ironically, owing to the popularity of reclaimed materials (particularly for their 'instant authenticity'), prices can actually be higher than new versions.

www.uk-reclamation.co.uk www.salvoweb.com

## RAINWATER HARVESTING

It is always better to reduce waste before looking for new sources – so prior to considering rainwater harvesting methods, you should first reduce your water wastage as much as possible. However, in our rainy country it seems to make sense to collect rainwater for household use before it goes down the drain. This has a threefold advantage. Firstly, it means that less water is taken from reservoirs and rivers. Secondly, it means that rainwater run-off, causing flash flooding, is slightly reduced, and thirdly, it can be collected in situ in a relatively pure state. Although it is possible to filter rainwater to provide drinking water, it is better to use it for toilet flushing, garden watering and other non-potable uses.

The simplest way to make use of rainwater is to put a water butt on all downpipes, and save the water for use in the garden. This is a very cost-effective solution, as the equipment needed will cost only tens of pounds rather than the thousands you would need to spend on a more complicated system.

Water for toilet flushing needs no treatment, though the rainwater store should be covered to prevent the ingress of sunlight and animals. There are now several manufacturers offering packaged rainwater collection, storage and pump units. At present these are quite expensive (approximately £2,000), and you may be better off just carrying out water-efficiency measures.

How much water can be collected? An average house in the UK will have around 100m$^3$ of water per year running off its roof. As the average household use for toilet flushing is 35-70m$^3$ per year (depending on the toilet's flushing efficiency) it is possible, with enough storage, to save this amount of water – worth perhaps £70-£150 per year.

The water from the roof will need to be filtered to prevent dirt, leaves and bird droppings entering and contaminating the flushing system. The best filters (e.g. WISY system at 90% efficiency), operate by surface tension, are self-cleaning, and automatically reject the 'first flush' of contaminated rainwater. Between showers they dry out, so contaminant organisms do not survive. Storage tanks should be sized to contain 1m$^3$ of water per 30m$^2$ of roof. It is preferable to have the storage tank underground, both for aesthetic reasons and as protection against frost. An automatic pumping system is required to deliver water to the WCs. There will normally have to be a mains 'back-up' which can be arranged to come on automatically, and which has to be fitted with approved air gaps to prevent cross-contamination.

PASSIVHAUS DWELLINGS USE AN EXTENSIVE RANGE OF LOW-ENERGY FEATURES AND PRINCIPLES TO ACHIEVE ULTRA-LOW CARBON EMISSIONS

WISY system – Rainharvesting Systems Ltd: 0845 223 5430
www.rainharvesting.co.uk Freerain: 01636 894906 www.freerain.co.uk
REWATEC: 01844 238111 www.rewatec.co.uk

## RAMMED EARTH

Rammed earth, or 'pisé de terre', is an ancient and worldwide form of wall construction that is blossoming today. In China, Yemen, Nepal, Egypt and Morocco the technique has been used for millennia; the Roman armies introduced rammed earth building to the south-east of France. In central Europe, earth building flourished in the 19th century and was revived after the first and second World Wars – in the UK by Portmeirion architect Clough Williams-Ellis.

The tallest earth building in Europe is in Weilburg, Germany. Built in 1828, it has seven storeys and rammed earth walls tapering from 750mm to 400mm thick. A whole village – Domaine de Terre – was built of earth in the 1980s. In Australia, there is today a thriving rammed earth construction industry building hotels and houses.

So why are an increasing number of self-builders with eco principles in mind interested in using it? Not only is it sustainable, thanks to its obvious natural occurrences, but new research is also beginning to emphasise its qualities as a solid form of construction: think of rammed earth as 'instant rock'. The natural processes that make sedimentary rock over thousands of years are imitated by the earth rammer. Rammed earth construction consists of moist, loose subsoil highly compacted between shuttering, in layers of 100-150mm depth. The mechanical compaction compresses this depth to about half, and forces the clay molecules to bond with the various aggregates. The shuttering is then moved along or upwards to form a whole wall. Because of the relatively dry mix, the shrinkage of rammed earth elements is much less than for other earth-building methods, and the strength is correspondingly higher. A rammed earth wall will dry out and become as tough and beautiful as sandstone, as long as it is protected from damp.

THINK OF RAMMED EARTH AS 'INSTANT ROCK'. THE NATURAL PROCESSES THAT MAKE SEDIMENTARY ROCK OVER THOUSANDS OF YEARS ARE IMITATED BY THE EARTH RAMMER

Knowing the exact composition of the soil and the correct amount of water to be added is critical for the success of this method. Usually the builder will take the material found on site and adjust its composition by mixing in other 'bought-in' materials. Where there is no suitable material on site, a mix can be designed and assembled from available quarry wastes and clays.

At the Centre for Alternative Technology, tests were carried out on various samples of local quarrying 'overburden' to determine the composition and strength of the material. The best samples showed a compressive strength of 2.29 newtons/mm$^2$ — well in excess of what is needed to support a lightweight, two-storey building.

Where stabilisers such as lime or cement are considered necessary, these should be kept to a minimum. Where they are used routinely, as with most modern earth construction in Australia, some of the environmental benefits are reduced.

Earth mixes can be compressed in a block-making machine. Blocks are produced in standard sizes and allowed to dry, under cover, for several weeks. They can then be laid in a lime- or clay-based mortar and rendered with the same. Stabilised earth blocks are made stronger and more durable by the addition of small amounts of lime or cement (5-10%).

While rammed earth construction has not yet reached the commercial scale found in Australia, much useful research has been carried out in universities, and there is now a DTI handbook, signifying its growing popularity.

In Situ Rammed Earth Co Ltd: 020 7241 4684 www.rammed-earth.info

## REED BED WASTE WATER TREATMENT

If your site is off mains drainage, there is an eco option for waste treatment. Rather than using an electrically driven treatment plant, you can opt for a more natural approach by constructing a reed bed filtering system. You do need a fair amount of space – about the size of half a tennis court – and you ideally need a ditch or an outlet of some sort to take the cleaned waste water. Reed beds are normally used in conjunction with a septic tank which acts as a reservoir for the foul waste as it exits the drain system. The outflow enters the reed bed and slowly filters through it, being cleaned by the vegetation along the way.

Cress Water: 01884 839000 www.cresswater.co.uk
Rockbourne Environmental: 01202 480980 www.rockbourne.net YES (Yorkshire Ecological Solutions) Reedbeds: 0113 252 4786 www.yes-reedbeds.co.uk

## SAP RATING

The amendments to Part L of the Building Regulations in 2002 introduced the requirement for all new homes (including conversions, but excluding renovations) to be assessed as to their energy-efficiency standard, based on predicted heating and hot-water costs. This Standard Assessment Procedure (SAP) rating is calculated at the drawing stage and is measured between 1 and 120. One hundred is a home that is super efficient, while 30 is considered the low benchmark for a home that is inefficient. Your SAP rating can be calculated off your houseplans, and is influenced by such factors as insulation levels, type of boiler and heating system.

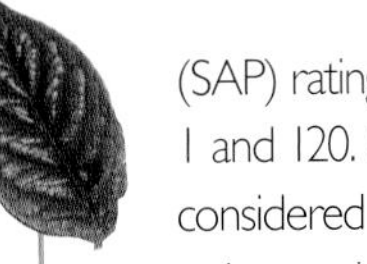

## SOLAR PANELS (THERMAL)

Solar thermal or hot-water panels are the best known of the microgeneration technologies, having been around the housing scene since the 1970s. The basic idea behind them is very simple: you place something very much like a radiator on your roof and the water circulating through the panel heats up when the sun is shining. This can generate a lot of hot water in summer time, but output falls by 90% or more in the winter months. Consequently, solar thermal panels are used

primarily to provide domestic hot water for the summer months and save on having to use a boiler for this task. They are also popular for use with swimming pools.

A more recent innovation is the development of evacuated tubes. These are more expensive than the traditional flat-plate panels, but they are more effective at harvesting energy from daylight, especially in the winter months. Whichever system you plump for, it is vital to assess your likely hot-water requirements so that you install the optimal area of panels. Unlike with PV solar panels, there is no way of exporting your surplus hot water, so it doesn't make sense to fit too large an area of panels.

Grants are available for the installation of solar thermal – see the section on Grants for more details and how they vary from country to country. A well-designed system could produce as much as 3,000kWh of usable hot water each year (cost in heating oil = £100) so there is reasonable hope of there being a sensible economic payback. Be warned that there are a number of sharp operators in this market who go about justifying charging double the normal installation costs (£7,000 instead of £3,500) on the back of over-inflated claims, such as 'cutting your heating bills in half'.

Powertech Solar: 01202 890234 www.solar.org.uk
Solar UK: 01892 526368 www.solaruk.net
Southern Solar: 0845 456 9474
www.southernsolar.co.uk
First Light Energy: 01732 783534
www.firstlightenergy.com
Viessmann: 01952 675060 www.viessmann.co.uk
Worcester Bosch: 0845 725 6206
www.worcester-bosch.co.uk
For more suppliers see Solar Trade Association:
www.greenenergy.org.uk

YOUR SAP RATING CAN BE CALCULATED OFF YOUR HOUSEPLANS, AND IS INFLUENCED BY SUCH FACTORS AS INSULATION LEVELS, TYPE OF BOILER AND HEATING SYSTEM

## SOLAR GAIN See Orientation

## STRAW BALES

In these days of increasingly technological solutions to the question of how to build, it is unusual – and refreshing – to find a simple, plant-based building material that can provide the structure and/or the insulating infill walls for several types of low-rise buildings.

Straw is a renewable raw material that is a by-product of grain production, and in many arable areas is a waste product that is not easy to dispose of – meaning straw bales are low cost and easily available.

Walls made with straw bales can be fully load-bearing or the bales can be used as an infill between timber frames. In either case, the bales should be as tightly compacted as possible. A strip foundation and plinth wall up to DPC level is laid in the normal way but 450mm wide, and bales are then stacked like giant building blocks, with staggered joints. Each course is pinned to the one below with timber or metal rods. Door and window openings are sized as multiples of bale lengths (900mm) or half lengths and lined with framed plywood. Timber plugs or wedges for wall fixings and sockets can be hammered in as the walls go up. The bales are then rendered inside and out, ideally with a lime render or clay plaster. Roof construction is conventional, with the rafters or trusses supported on an extra-wide wall plate.

One of the main advantages of straw bale construction is that you will get a highly insulated wall simply by virtue of the bale widths and, therefore, the wall thickness. Typical U-values quoted are from 0.21-0.13W/m²K. Straw bale walls also provide relatively good sound insulation and, contrary to popular belief, are very good at resisting the spread of fire. It is, in fact, surprisingly difficult to set fire to a well-compacted bale, and the render or plaster coat will provide additional protection. Small rodents can be kept out of what might be a tempting nesting site by keeping the bales well clear of the ground, with physical barriers installed if necessary.

The only real problem in building with straw – but it can be a big one – is keeping the bales dry at all times. Walls under construction must be well protected from moisture. In a framed building the roof should go on first and the walls be built up underneath it. If any bales do get wet they should be replaced immediately or else they will start to compost.

Good design should help to keep the bales dry throughout the life of the building – providing overhanging eaves and a generous distance between ground level and DPC (150mm is the minimum, but 200-300mm is better) will avoid water penetration and splashback.

The Straw Bale Building Association: 01442 825421
www.strawbalebuildingassociation.org.uk

## STRUCTURAL INSULATED PANELS (SIPS)

Structural insulated panels (SIPs) are seen as an inherently eco-friendly way of building, as manufacturers claim they can achieve U-values of as low as 0.22W/m²K, combined with quick installation times.

SIPs building is sometimes referred to as timber frame without the timber. Like panelised timber frame, it works on the idea of building up wall and roof panels on flat surfaces and then hoisting them into position, in contrast with masonry build systems which build walls

up in situ. Where SIPs differ from timber frame is that they gain their strength not from any timber skeleton, but from the rigidity of the panels themselves. Structural insulated panels are essentially a sandwich: the filling is a solid thickness insulation, and the bread is made of rigid building boards such as plywood or orientated strand board (OSB). These layers are bonded together which has the effect of making the panels extremely robust. In many ways, the technique is similar to how aircraft wings are designed, with two skins wrapped around a lightweight core and then welded together to form a single element.

The advantages are that you create a structure with superb insulation levels, few cold bridges and excellent airtightness levels. Plus, in the right hands, SIPs have the potential for faster construction speeds than timber frame. The panelised roof elements lend themselves to building rooms in the roof, in a simpler and quicker manner than traditional methods.

The main attraction of SIPs building is that it provides a fantastic insulation level built into the fabric. This is particularly attractive for sloping roofs, which have become awkward to insulate using traditional methods. SIPs offer exceptionally good airtightness levels, an increasingly important aspect of low-energy housebuilding.

Custom Homes: 01787 377388 www.customhomes.co.uk
SIPs Industries: 01383 823995 www.sipsindustries.com
SIP Build Ltd: 0870 850 2264 www.sipbuildltd.com
Build It Green Ltd: 0870 2000 358 www.buildit-green.co.uk
Siptec: 01234 881280 www.siptec.co.uk
Kingspan Tek: 0870 850 8555 www.tek.kingspan.com/uk
Enviropanel: 0870 284 6244 www.enviropanel.com
SIPs Association: www.sips.org

WALLS MADE WITH STRAW BALES CAN BE FULLY LOAD-BEARING OR THE BALES CAN BE USED AS AN INFILL BETWEEN TIMBER FRAMES

## THERMAL MASS

Thermal mass is a topic of hot dispute in the green building movement. Advocates suggest that by using thermal mass correctly in a house or, indeed, any building, you can greatly reduce the winter heating load and also reduce the need for summer cooling.

Critics point out that it all depends on how the building is occupied and that in many instances it can actually make matters worse in both summer and winter.

The theory behind thermal mass is very similar to that used for storage heating. By using lots of heavyweight material in a house such as concrete, brick or tile, you create a heat reservoir, as this material takes a lot of energy to heat up. If you get the design right, you can charge the heat reservoir with sunshine during the day and then enjoy the radiated heat coming from the brick and concrete through the hours of darkness. This essentially is the principle of passive solar design. Ideally, at least some of the high-mass elements need to be exposed to direct sunlight so that they capture the solar radiation; but this doesn't always happen. The downside of thermal mass is felt when the house is left unoccupied for any length of time in winter, then the heat gradually dissipates and it takes a long time (and a lot of heat energy) to get all that thermal mass charged up again.

Different principles are at work during the summer months when thermal mass is said to help keep homes cool, without the need for air conditioning. High-thermal-mass homes do tend to stay a little bit cooler in the daytime because they absorb some of the heat and this means that they even out the temperature fluctuations between day and night. However, even here there is a counter viewpoint that suggests that the key to maintaining comfortable homes in a heat wave is more to do with ventilation than thermal mass and that the one time when residents appreciate cooler indoor temperatures is at night, which is when high-thermal-mass houses tend to be hotter.

It's a complex debate with each and every point disputed. Though there is a hidden agenda to be aware of: behind the debate lies a turf war between advocates of masonry and other heavyweight construction systems on the one hand, and timber and steel frame on the other.

## TIMBER

Wood is one of the most emotive of building materials and, if bought from the right sources, can also be one of the most environmentally friendly. Each cubic metre of wood stores around 0.8 tonnes of $CO^2$, and as wood sourced from sustainable forests is replaced, this represents a net gain.

It is important to buy timber that is from well-managed sources – softwood is generally not a problem as the vast majority of what we buy in the UK is from European forests, which are all under effective Government regulation. Hardwood, however, usually comes from tropical or sub-tropical forests, and therefore it is important to ensure it is sourced from a certified forest. The UK Government recognises five schemes: FSC (Forest Stewardship Council), PEFC (Programme for the Endorsement of Forest Certification), CSA (Canadian Standard Association), SFI (Sustainable Forest Initiative) and MTCC (Malaysian Timber Certification Scheme) – look out for their accreditation on timber you intend to buy.

www.woodforgood.com. For a list of timber frame suppliers visit www.homebuilding.co.uk or the UK Timber Frame Association at www.timber-frame.org

## THERMAL STORES

Thermal stores are sometimes used in preference to more conventional hot-water cylinders. They are similar in many respects: tanks of hot water are used to distribute heat around the house and to provide hot water to the taps. There is an energy storage potential here that can be very useful to combine with green heating systems, especially solar hot-water panels, which only produce hot water in daylight hours. In effect, a thermal store also acts as a hot-water battery. The development of thermal stores is something that is still continuing and the latest versions have the potential to drive at least some of the house space-heating requirements as well, just by harnessing the hot water from solar panels.

DPS Heatbank: 01372 803675 www.dedicatedpressuresystems.co.uk
Chelmer Heating: 01245 471111 www.chelmerheating.co.uk

## U-VALUES

A measure of an object's heat transmission due to thermal conductivity. The lower the U-value, the more restrictive the material at allowing heat to pass. Therefore, a wall with a U-value of 0.2 is considered to be more efficient than a wall with a U-value of 0.3. It is indicated in units of watts per metre square per degree kelvin ($W/m^2K$), where kelvin is identical to degrees Celsius. The Building Regulations require external walls of houses to have a U-value of $0.3W/m^2K$.

## UNDERFLOOR HEATING (UFH)

This is the process of heating a home not through wall-mounted radiators, but by running pipes underneath the floor surface. In solid floors the pipes are set in screed, while in suspended floors they are clipped to the floor deck, in between the joists. UFH is seen as a green option because, owing to the greater surface area that is warmed up as opposed to radiators, it can operate at temperatures of around 50-60°C, as opposed to the 80°C required by conventional systems. As a result, it is perfectly suited to solar and geothermal energy sources. Warm-water underfloor heating (electric versions are available and run on low energy requirements and non-invasive thin matting and wiring, making them particularly suitable for retrofit installations) costs in the region of £2,000 for a modest-sized ground floor.

For a full list of underfloor heating suppliers visit www.homebuilding.co.uk or visit the Underfloor Heating Manufacturer's Association: www.uhma.org.uk

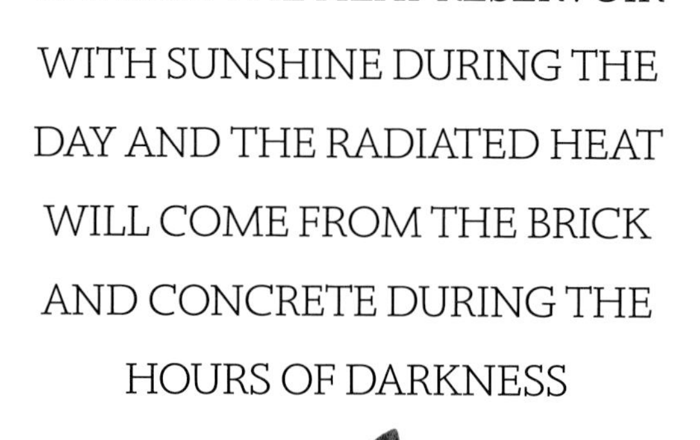

## VENTILATION

Put simply, ventilation is the removal of 'stale' air from inside a building and its replacement with 'fresh' air from outside. Adequate ventilation is essential in your home to maintain a healthy environment but also to prevent the build-up of excess levels of humidity and to provide air for fuel-burning appliances. In addition, a well-designed ventilation system will minimise heat loss from your home and, potentially, reuse warm air for heating, making it an essential requirement of a green home.

Buildings are ventilated through a combination of air infiltration and purpose-provided ventilation. Infiltration is the uncontrollable air exchange between inside and outside a building through a range of air-leakage paths in the building structure, i.e. drafts. Purpose-provided ventilation is the controllable air exchange by means of a range of natural and/or mechanical devices such as trickle vents in windows and electric extract fans.

The recent changes to the Building Regulations place more importance than ever on ventilation. This is because ventilation is rapidly becoming the greatest source of heat loss in new dwellings as the standard of insulation continues to improve. The revised Part L, the regulation dealing with energy efficiency, places a great emphasis on making buildings as airtight as possible and air-pressure-leakage testing has become a mandatory requirement for all dwellings, to measure air permeability and show any unacceptable leakage. This in turn places a greater importance on controlled, purpose-provided ventilation as opposed to infiltration.

The least costly option is probably going to be individual extract fans in wet areas and passive vents (trickle ventilators or airbricks with sliding hit-and-miss vents), or a positive input ventilation system (PIV). If you are building an eco house with a very low heat requirement then your choice is likely to lie between a passive stack system that uses no direct electricity, a PIV system, or a heat-recovery ventilation system (HRV). If you want to be even more ecological, you could opt for a mechanical system that preheats the incoming fresh air via solar panels: when the heat is not required it is diverted to the hot-water cylinder.

Heat-recovery ventilation systems have a central fan unit which draws stale air from wet areas and balances it with fresh air drawn from outside which is filtered and pre-warmed via a heat exchanger (and sometimes an electric element) before being blown into dry areas.

HRV systems are designed to continuously extract polluted air from moisture- and odour-producing areas such as kitchens and bathrooms.

This is done at a low background level with an occasional boost when required. The air is carried via a network of ducts concealed within the building structure, connected to a central mechanical fan unit, usually located in the loft or other out-of-the-way location. Before filtered air from the outside is supplied to 'dry' habitable rooms (bedrooms, living rooms etc), heat is transferred from the stale air via a heat exchanger in the power unit, with a 70-93% efficiency.

The relative cost-effectiveness of heat recovery will depend on its efficiency level, i.e. how much of the energy consumed for space heating is recycled. In a well-insulated and fully airtight house, the efficiency will be high – whereas in a house with lots of leaks and full of trickle vents and airbricks, its efficiency will be seriously compromised. This is why the Building Regulations have become much stricter about airtightness in new buildings.

Starkey Systems: 01905 611041 www.centralvacuums.co.uk
Villavent: 01993 778481 www.villavent.co.uk Passivent: 0161 962 7113
www.passivent.com Vent-Axia: 01293 530202 www.vent-axia.com,
ADM: 01756 701051 www.admsystems.co.uk AllergyPlus: 0870 190 0022
www.allergyplus.co.uk Brook Airchanger: 028 9061 6505
www.brookvent.co.uk Unico: 01443 684828
www.unicosystem.co.uk

## VOLATILE ORGANIC COMPOUNDS (VOCS)

Chemical compounds such as hydrocarbons and ketones that are commonly found in artificially produced building materials – e.g. carpet backing, paints, solvents. They are usually the reason behind that 'new house smell'. They can, however, potentially cause headaches, nausea and contribute to Sick Building Syndrome.

THE COST-EFFECTIVENESS OF HEAT RECOVERY WILL DEPEND ON ITS EFFICIENCY LEVEL, I.E. HOW MUCH OF THE ENERGY CONSUMED FOR SPACE HEATING IS RECYCLED

## WIND TURBINES

The use of wind turbines to generate electricity is a rapidly developing technology. We are now all familiar with the giant 100-metre-high wind farm turbines, but there are developments at the smaller scale as well, many of them aimed at individual householders. If you want to generate renewable electricity, and you have the space, then a pole-mounted wind turbine is probably the most attractive option.

However, the output of wind turbines is determined by a) their size and b) their height above ground.

The giant 2MW wind farm turbines produce a lot of electricity; the medium-sized 6kW garden turbines produce a reasonable amount; the tiny roof-mounted ones (rated at around 1kW) produce very little indeed and despite lots of interesting developments in this field over the past two years, it appears unlikely that they ever will. Our advice at present is to steer well clear of roof-mounted wind turbines and look instead at erecting a wind turbine on a suitable pole away from the house. The success or otherwise of the scheme is very dependent on the local topography, the height of the turbine above ground, the size of the turbine and the local wind speeds. They are really only suitable in exposed locations.

But with the Government's move towards zero-carbon homes, there is a huge interest in wind turbines to complement the PV cells on the rooftops. Grants are available for wind turbine installations – see the section on Grants to find out more.

British Wind Energy Association: www.bwea.org
Proven Energy (small wind turbines): www.provenenergy.com
Solar Energy Alliance: www.solarenergyalliance.com

## ZERO-CARBON HOMES

A home that, over the course of a year, doesn't add any $CO^2$ to the atmosphere. The Government recently announced that it wants all new houses in England to be zero carbon by 2016. In Wales, the intention is to do this by 2011. As a typical home in Britain releases between 10 and 20 tons of $CO^2$ each year, to make a home zero carbon, a two-pronged approach is required: reducing energy requirements (eg through better insulation) and also generating energy on site (eg through solar panels). The Government's definition of zero carbon does not account for the carbon impact embedded in the materials used in construction.

The Government has published a document called Building a Greener Future: Towards Zero-Carbon Development which outlines the way

housebuilders (including, of course, self-builders) need to move towards greater energy efficiency. Its proposal is:

- 2010 sees a 25% improvement in the energy performance required under the Building Regulations.
- 2013 sees a 44% improvement.
- 2016 sees every new home built as zero carbon.

The Government's Code for Sustainable Homes, published in December 2006, is designed to give guidance to housebuilders who want to exceed the existing Building Regulations targets. Effectively, it is a manual for building zero-carbon homes. The Government in essence intends the (voluntary) Code to be a precursor to significant upgrades in the (compulsory) Building Regulations.

For a full analysis of Zero-Carbon Homes, see the July 2007 issue of *Homebuilding & Renovating Magazine*

DETACHED HOUSE ON ONE FLOOR PLUS MEZZANINE £300,000

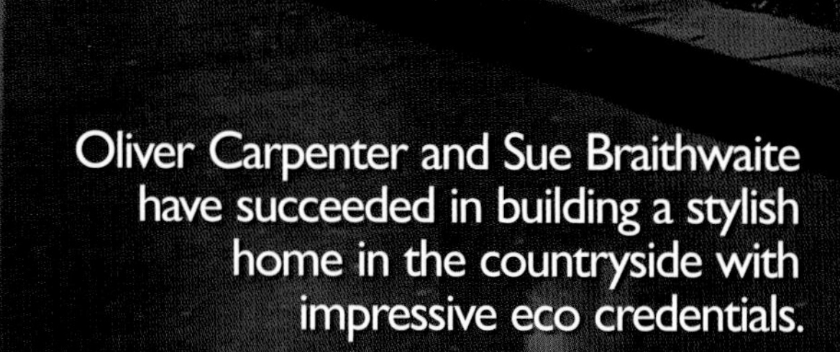

Oliver Carpenter and Sue Braithwaite have succeeded in building a stylish home in the countryside with impressive eco credentials.

# BLENDING IN

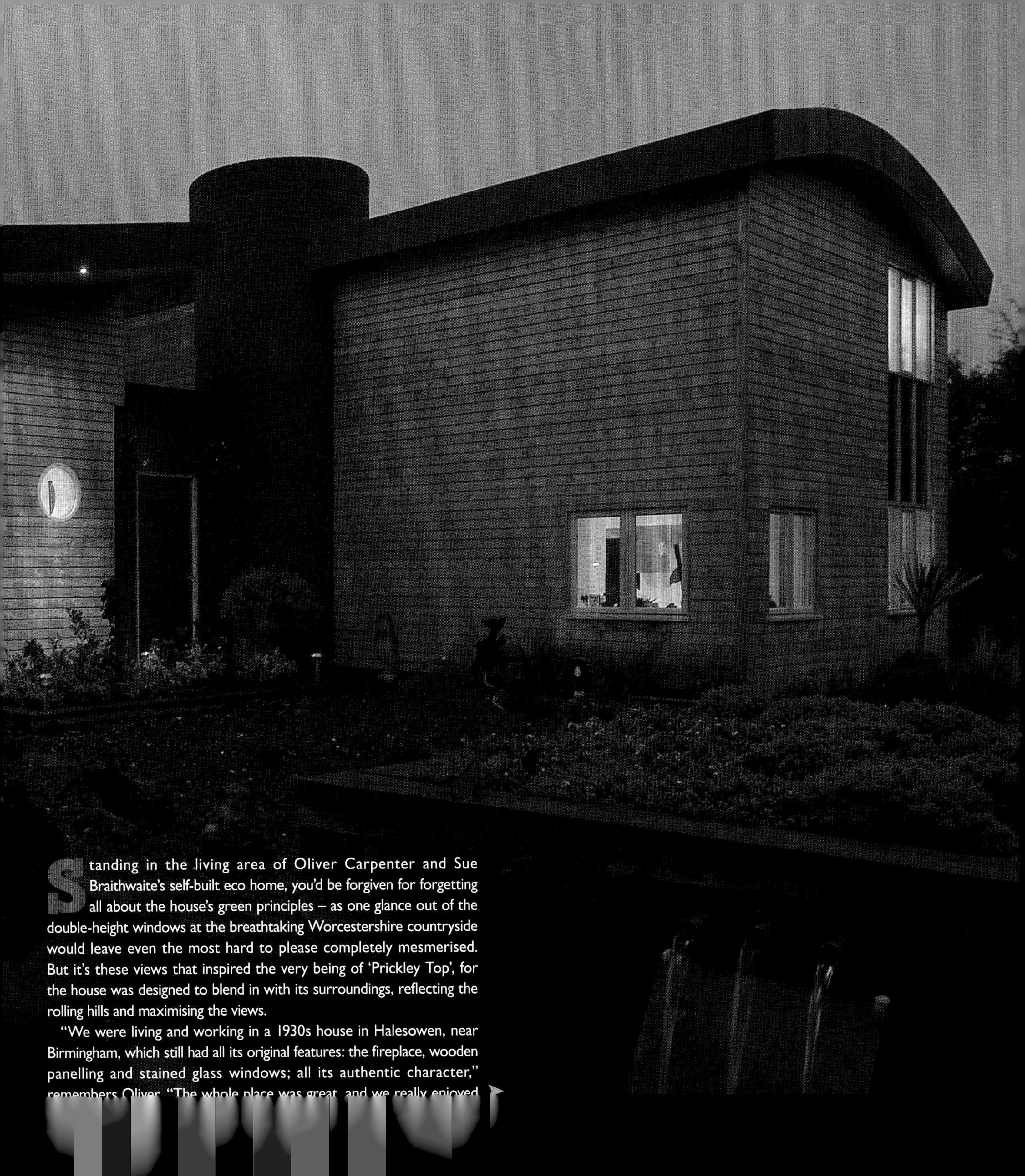

Standing in the living area of Oliver Carpenter and Sue Braithwaite's self-built eco home, you'd be forgiven for forgetting all about the house's green principles – as one glance out of the double-height windows at the breathtaking Worcestershire countryside would leave even the most hard to please completely mesmerised. But it's these views that inspired the very being of 'Prickley Top', for the house was designed to blend in with its surroundings, reflecting the rolling hills and maximising the views.

"We were living and working in a 1930s house in Halesowen, near Birmingham, which still had all its original features: the fireplace, wooden panelling and stained glass windows; all its authentic character," remembers Oliver. "The whole place was great, and we really enjoyed

## "WE WERE TRYING TO BUILD A GREEN HOUSE AND THE PLANNERS WERE DOING THEIR BEST TO GET IN THE WAY – IT DIDN'T MAKE SENSE"

living there; but we hankered after a house that was designed with us in mind, instead of those who were there before."

It was not only the idea of a home designed to their own needs that appealed to the couple. "We had been developing more of an interest in green issues, such as buying local, organic produce and we liked working from home, which is in itself an ecological principle, so it struck us that any house we built should be as green as possible – or at least as green as we could afford – though we knew nothing about building, green or otherwise," continues Oliver.

"We drew a circle around where we were prepared to live and began a search for land." After trying all the usual methods, such as dropping leaflets through doors of empty buildings and pestering agents, the couple finally came across a tired 1930s urban bungalow right in the middle of the perfect plot. "It faced the wrong way and didn't do justice to its surroundings – it was the wrong house in the wrong place," states Oliver. However, it wasn't until the structural engineer's report declared the property unsound that Oliver and Sue realised they had to knock it down and start again.

"Though we were planning a green build project, the first thing we did was demolish a house – which was not a good start!" laughs Oliver. "So we recycled everything we could: the roof tiles went to a local farmer; the windows, doors and woodwork went off to reclamation places; the unused central heating system and all the pipes and radiators found new homes."

When it came to design, the couple's main goal was to blend in with the environment. "We feel that most houses don't relate to the land they sit on. We couldn't understand that. We didn't have a design in our heads – we were open-minded."

Oliver went to see seven different architects, some AECB (Association for Environment Conscious Building) registered; some not. The seven were shortlisted to four, who were then invited to the plot to discuss the couple's ideas. From the meetings, Neill Lewis, a one-man practice from Malvern with strong green credentials, was chosen. "We felt he was someone we could work with throughout the project," explains Oliver. "Obviously there are always going to be ups and downs, so we picked him for the relationship that we could have with him, as much as anything else. We explained about the eco features we wanted to include and showed him pictures of buildings that we really appreciated, and his first sketch, externally, turned out to be pretty much the final design."

However, settling on a design was the easy part – getting planning ➤

**The brick tower, the main focus of the open-plan ground floor, hides a spiral staircase to the mezzanine bedroom and roof.**

**The film-set-style lighting rig allows flexibility in the interior scheme.**

"WE SHOULD BE ENSURING EVERYONE INSULATES EFFECTIVELY... THIS IS EASY TO DO – WE NEED TO SORT OUT THE SMALLER ISSUES BEFORE THE BIGGER ONES"

permission was not to go so smoothly. "The planners were obsessed with fitting buildings into the local vernacular," says Oliver. "They let people build houses that look like all the other houses – and ours didn't. So they kept delaying the process by asking questions like, 'What will the cladding look like in five years' time?' We were trying to build a green house, and though they are a green/liberal council, they were doing their best to get in the way – it didn't make sense." In the end, the design was approved with a few minor amendments.

After a further six-month delay involving two great crested newts, who prohibited the build while they were in hibernation, building at last began in June 2004 – though, controversially, Oliver and Sue didn't choose a green builder. There were none close by, so a local builder called Roger Bloomfield was hired, along with a team the couple felt they could "weather the process with." Optimistically, Oliver is now of the view that they have perhaps "added to the amount of eco builders out there."

As one might expect, Prickley Top utilises eco-friendly features wherever possible. The sedum roof was the biggest splurge at a cost of around £25,000, though it is justified by Oliver: "It is such an amazing feature and an inherent part of the design. The 'rolling hill' shape wouldn't have worked without a growing roof – not only does it provide extra insulation, but in a way we're replacing the footprint of the land we built on."

All wood used is FSC and UK grown, the window systems being the only exception, which were brought in from Denmark via EcoMerchant. "Though they have travelled a long way, the company is very eco-aware," Oliver explains. All the gravel came from a quarry five miles away and the slate came from Wales. Oliver is a big believer in using local materials wherever possible: "The thought of boats travelling the world, laden with stone and wood is just crazy! It's bad news for the planet and global warming – it's not like we haven't got our own."

The house is well insulated with Warmcel recycled newspaper insulation throughout. "The effects are just fantastic," enthuses Oliver. "We have underfloor heating, and the warmth doesn't escape."

In addition, the couple have installed an ultra-efficient woodburning fire, dual-flush toilets and non-power showers. The kitchen units are recycled 1960s English Rose units, which have had their original enamel stripped, and add a quirky character, and the fridge is also recycled. All the appliances are A-rated – though the couple admit to one sin: they do have a dishwasher.

The one thing missing from Prickley Top, and a big disappointment for Oliver and Sue, is that it does not collect its own rainwater. "We're one of the few new eco houses that doesn't," sighs Oliver. "We had planned and ➤

**The enormous double-height windows ensure the house is not short of passive solar gain.**

"THE SURROUNDINGS ARE LOVELY AND LIVING HERE HAS EXCEEDED OUR EXPECTATIONS"

costed it in, but the green roof installer told us there wouldn't be enough water run-off. But now, of course, we have discovered there is lots of water run-off. We have made a big mistake, basically through misinformation. One of our criteria from the start was not to have visible drainpipes, so now we risk ruining the line of the building to put in water butts. It's a real shame."

The obligatory solar panels are also a bit of a sore point, as the couple were denied their ClearSkies grant because the supplier changed the installer without informing the organisation, and Oliver and Sue lost out on £400.

Just because Prickley Top is a near-exemplar green house, it doesn't mean that style and design ingenuity have been forgotten. One of the most exciting features is the glass-covered integral well, which appears bottomless, until Oliver flashes a light down and you can just about make out some water in the depths below. In an inversion of the well, there is a narrow tower standing opposite. This tower contains LED-lit stairs leading to the mezzanine bedroom, rooftop and all-important storage space.

The double-height, open plan sitting/dining/kitchen space is dramatic and not short of passive solar gain. The walls, ceilings and floors are nearly all neutral, but Oliver and Sue have introduced some bold colours too, namely in the pink-floored kitchen.

**A narrow tower opposite the glass-covered integral well displays stylish design ingenuity.**

It's not just the house that's had the green treatment: 12,000 trees have been planted; four wildlife ponds have been built; and the couple are working on a one-acre wildlife meadow.

Oliver and Sue hope their house can show people that, while it is not easy to be green, it can be done by people with no experience and without an unlimited budget. "One of the dangers is that, in the past, green building was seen as other-worldly," says Oliver. "Now, just as green requirements have become mainstream, the only people making the news are talking about features like wind turbines and heat pumps, which are way out of reach in terms of price and practicality for most people. Instead we should ensure everyone insulates effectively, and installs dual-flush toilets and ultra-efficient windows. This is easy to do – we need to sort out the smaller issues before the bigger ones, like personal power generation."

Oliver and Sue are thrilled with the finished house. "We came into the project without knowing anything about building and as a result we've had to compromise, but the place is fantastic. The surroundings are lovely and living here has exceeded our expectations. We still can't quite work out how we ended up here — we really are very privileged." ●

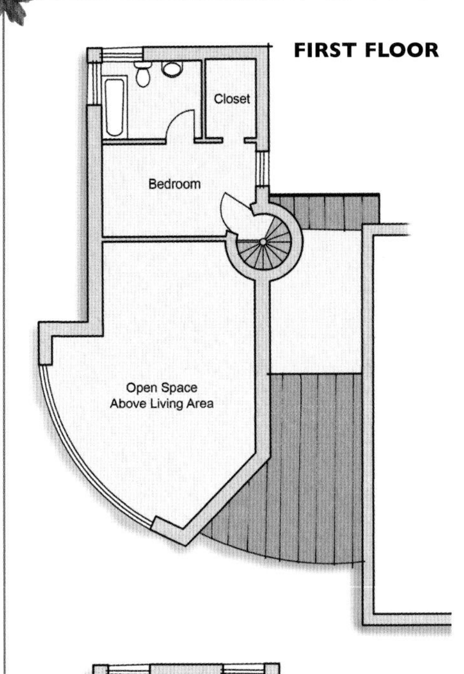


## FACT FILE

**Names:** Oliver Carpenter and Sue Braithwaite
**Professions:** Marketing consultant and management consultant
**Area:** Worcestershire
**House type:** Three-bedroom eco-friendly detached house; one floor plus mezzanine
**House size:** 230m²
**Build route:** Self-managed, with architect and builder
**Finance:** Private plus Ecology Building Society mortgage
**Construction:** Brick and block, steel girders in roof, timber cladding
**Build time:** June 2004-May 2006
**Land cost:** £230,000
**Build cost:** £300,000
**Total cost:** £530,000
**Current value:** Unknown
**Cost/m2:** £1,304

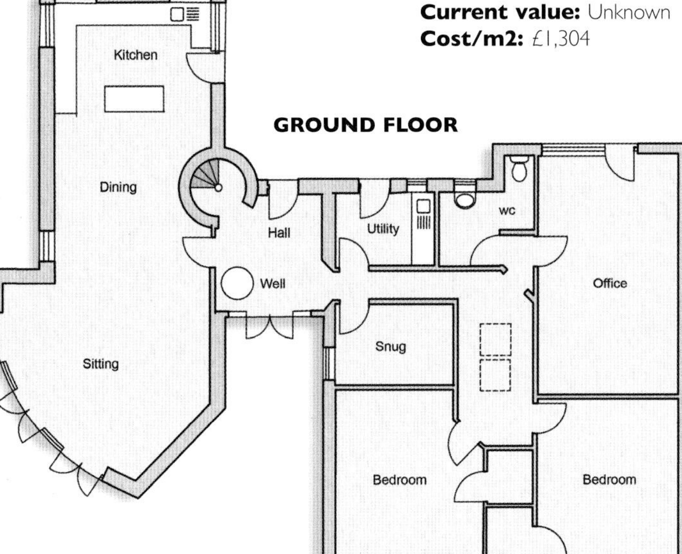


## FLOORPLAN

**The ground floor layout is based around a large open-plan living kitchen area, making the most of the large amounts of south-facing glazing to warm through the house. The bedroom accommodation is kept in a separate wing of the ground floor in addition to upstairs.**

## USEFUL CONTACTS

**Architect** Neill Lewis: 01684 563356 **Builder** Roger Bloomfield: 07737 470022 **Mortgage** The Ecology Building Society: 0845 674 5566 www.ecology.co.uk **Advice** Association for Environment Conscious Building: 0845 4569773 www.aecb.net **Bespoke Danish wooden windows** Vrogum A/S: +45 76541111 www.vrogum.dk **Planted roof** Bauder Ltd: 01473 257671 www.bauder.co.uk **Warmcel insulation** Stewart Energy Conservation: 020 8648 6601 **Western red cedar cladding** Stourhead: 01747 840643 www.stourhead.com **Low-energy sustainable underfloor heating, condensing boiler and solar panels** Eco Hometec: 01302 722266 www.eco-hometec.co.uk **Windows** Ecomerchant Ltd: 01795 530130 www.ecomerchant.co.uk **Oak flooring** Whitehall Reclamation www.white-hall.co.uk **Kitchen equipment** Stourbridge Kitchens: 01384 371201 **Stone spiral staircase** Blanc de Bierges: 01733 202566 **Woodburning fire** Continental Fires: 01694 724199 **Gas supply** Calor Gas: 0800 626626 **External slate** Berwyn Slate: 01978 861897 www.berwynslate.com **Aluminium kitchen units** Source Antiques: 01225 469200 www.source-antiques.co.uk **Pink Chesterfield** English By Design: 01384 390333

The new house is located on the site of a former sawmill, and the planners were happy to see any new design retain the basic form of the original building.

# HIGHLAND SETTLERS

**Artisans Eric and Meira Stockl have built a characterful new timber- frame home and workplace with strong ecological credentials.**

WORDS: CAROLINE EDNIE
PHOTOGRAPHY: ANDREW LEE

**The use of underfloor heating in the ground floor work area maximises available usable space and provides a more consistent temperature level.**

Eric and Meira Stockl have taken a great deal of pleasure in relocating from a 14-room mill conversion in Yorkshire to a bespoke bijou two-bedroom rural retreat in the North East Highlands. Yet, to describe the Stockls' new home as simply a worthy exercise in downscaling is not to tell the whole story. For in addition to their new cosy living arrangements, Eric, who is a practising potter and Meira, a weaver, have also literally created their very own cottage industry on the site.

Indeed, when the couple relocated to the Highlands three years ago to be near to their son Mark and two grandchildren, maximising workspace for Meira's looms and Eric's kiln and potter's wheel was uppermost in the couple's minds. As Meira explains, "We started off working in a tiny flat in the east end of London – Hackney before it got posh. Then we worked in an old warehouse space in Yorkshire, which had no plumbing or electricity." The couple were doggedly determined to finally work in comfort.

Fortuitously, the couple entrusted their simple vision for the site – which contained a derelict stone sawmill, a few outbuildings (one of which is now Eric's pottery studio), as well as great mounds of sawdust – to local Ullapool-based architect and builder Bernard Planterose of North Woods Construction. As a result the Stockls' new home/studio is as commodious, cosy, and eco-savvy as live-work spaces come.

Eric and Meira requested that the ground floor was to provide as big and light a studio for Meira as possible, and the headroom had to be 2.6m for ➤

"THE WEATHER WAS NOT ON OUR SIDE. WE STARTED IN NOVEMBER AND I'D NEVER DO THAT AGAIN..."

**The whole house was skim plastered and painted with an expensive 'Kayem' breathable paint which is silicate based rather than oil.**

the looms. The main living space therefore had to be on the first floor, and the couple also requested a balcony of some sort. Two bedrooms were asked for – one upstairs and one on ground level – and the studio required its own bathroom. The couple's final request was for a solid masonry chimney to accommodate a multi-fuel stove on each floor.

As well as designing the house, Planterose and North Woods Construction were also the main contractors. The project was a close collaboration with Nor Build Ltd, who not only fabricated the whole house in pre-insulated panels but also milled, kilned and machined every stick of timber on their own premises. They designed the panel system and the post and beam kit and were in charge of erecting the superstructure. On top of that they made all the units for the kitchen and elsewhere, and the worktops.

The house structure combines a post and beam spine with a normal stud framed exterior wall. "The basic timber post and beam structure of the house is a superb solution as it holds the first floor up, and a huge ridge beam holds the roof up," explains builder Bernard. "It's ideal for a house like this because it is open plan. It was engineered by John Talbott of Findhorn Engineering, who pioneered this form of construction in the houses he has built in the eco village of Findhorn. A full post and beam house is expensive, but Talbott came up with idea that if you put the posts and beams in the middle you get the impression of a post and beam house but without the expense. In a £100,000 build I think this adds around 10% to the cost."

In terms of planning, the project was extremely painless. Bernard attributes this to issuing a design statement outlining the whole design ethic and sustainability proposal to the planning department as a pre-planning exercise. The design proposal for the Stockls' new home was far from controversial, however, as it very much took on the old sawmill as its starting point, with its very straightforward rectilinear form and a steep black metal roof coming low to the ground on the north pitch.

The north elevation facing the road was kept as blank as possible, which was the best solution from an energy conservation viewpoint, as well as echoing the form of the old stone mill. The use of local timber also seemed entirely appropriate on a site which used to supply so much to the whole North West Highlands.

The construction process was not quite so straightforward, however, as

## "THE BASIC TIMBER POST AND BEAM STRUCTURE OF THE HOUSE IS A SUPERB SOLUTION AS IT HOLDS THE FIRST FLOOR UP, AND A HUGE RIDGE BEAM HOLDS THE ROOF UP"

Bernard explains. "The weather was not on our side. We started in November and I'd never do that again. Pre-insulated panels were built and constructed here using a small crane, but logistically it was a little bit dodgy, as the crane was too small – it was a jib on the back of a small lorry. This is something I would have done differently," he admits. "We also found that the only difficulty with post and beam is there are posts down the middle that take half the weight of the roof, so a lot of concrete goes under these posts. Incorporating these footings into the concrete slab foundations knocks the price up quite a bit too.

"The house is actually quite an achievement – on an ecological level it's faultless," claims Bernard. In addition to the almost entire use of indigenous timber throughout the exterior and interior, the walls are constructed using Warmcel insulation and Fermacell fibre reinforced plasterboard. "We used it because you can't really put ordinary plasterboard on panels that are being moved by crane – so we had to use a sheeting board that was far more resilient," explains Bernard. "Fermacell is highly ecological as it's made entirely from gypsum and unglued cellulose. It met all the objectives."

Meira cites the beautiful natural light which floods her ground floor workspace as being one of the loveliest qualities of the house. "Some things about the house are much nicer than we imagined, such as the light, and this is very important to me in terms of illuminating my loom when I am working," she explains.

Impeccable details distinguish the interior spaces throughout. "With the open plan kitchen we took standard units and hybridised them," explains Bernard. "We took off the unit doors and redid them, using the same Douglas fir that we've used throughout, and we replaced the worktops with Scottish sycamore. Similarly, with all of the doors, we bought cheap versions off the shelf, took the plywood out of the Scotch pine frames and, again, just used our own Douglas fir to create a horizontal emphasis to the design, which is reflected in the rest of the house.

"Many people have said as well as being very spacious, the house is actually quite spiritual," claims Bernard, a view very much shared by Eric and Meira. "We're really happy with the house," affirms Meira. "Not only do we have comfortable workspaces, but we also have a lovely open plan area upstairs to spend our leisure time. It is everything we hoped it would be (although some more storage space would be nice). It's a joy to live and work in."

### FACT FILE

**Names:** Eric and Meira Stockl
**Professions:** Potter and weaver
**Area:** Scottish Highlands
**House type:** Timber frame hybrid post and beam system
**House size:** 135m$^2$
**Build route:** Design and build contract
**Construction:** Masonry and steel with piled foundations
**Sap rating:** 100
**Finance:** Private
**Build time:** Sept '00-Aug '01
**Land cost:** £42,000
**Build cost:** £150,000
**Total cost:** £192,000
**House value:** £200,000
**cost/m$^2$:** £1,111

**4%**
**COST SAVING**

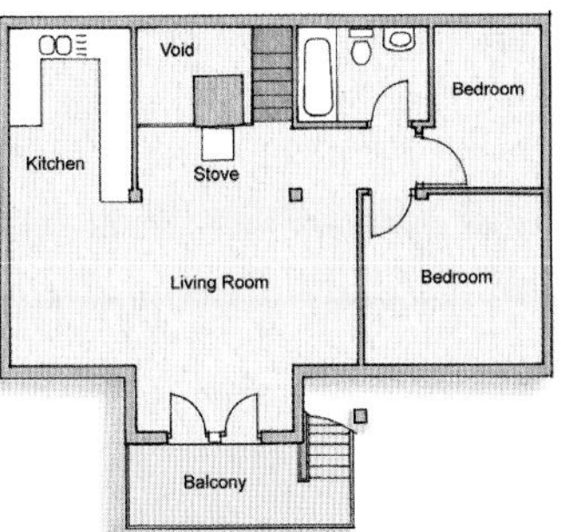


**GROUND FLOOR**

### FLOORPLAN

**The ground floor provides a large studio workplace with 2.6m ceilings. The main living space, kitchen and dining area are on the first floor. There are two bedrooms, one upstairs and one on ground level, and two bathrooms.**

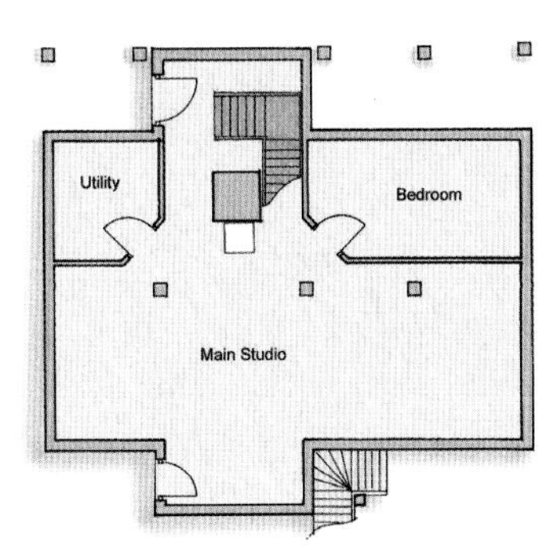


**FIRST FLOOR**

### USEFUL CONTACTS

**Designer** Bernard Planterose and North Woods Construction: 01854 613040; **Structural Engineer** John Talbott, Findhorn Engineering: 01309 690 154; **Main Contractor** North Woods and Nor Build Ltd: 01309 676 865; **Roofing** (Profiled metal sheet suppliers) Planwell Roofing: 01542 832170; **Windows** Treecraft Woodwork: 01862 810021; **Underfloor Heating** Invisible Heating Systems: 01854 613161; **Tiles** Tile Express: 01463 717860; **Insulation** Warmcel: 01495 350655; **Linings** Fermacell: 0870 609 0306; **Insulated Double Chimney** Isokern: 01202 861650.

# BACK TO NATURE

Martin Hutchins and Lucy Rutherford have built an unusual environmentally friendly house in Devon using local materials and labour.

WORDS: DEBBIE JEFFERY PHOTOGRAPHY: NIGEL RIGDEN

For Lucy Rutherford and her husband, Martin Hutchins, building their own house has been a life-changing experience in more ways than one. Situated in an Area of Outstanding Natural Beauty, Pebblebed House was designed to be a low-energy building with minimal impact on the environment, and the couple selected natural materials and designed the garden to encourage wildlife, as well as to grow their own organic vegetables. The change from their previous city life in Bristol couldn't be more pronounced.

"We planned to move to this part of Devon to be close to the Exe River – which we know very well," says Lucy, a former film sound editor, "and we lived in rented accommodation while we searched for a suitable house to renovate. When we found this quarter-of-an-acre site, there was already a small bungalow here, but we decided to try and replace it. We talked to the planners at length before we put

The curved roof enabled additional rooms to be incorporated at first-floor level and, together with the larch cladding, 'poppled battered' walls and buttresses, gives the house immense character.

**The kitchen from Alno (020 8898 4781) faces south-west to benefit from passive solar gain.**

"MOST BUILDERS SUCK THEIR TEETH WHEN YOU START TALKING ABOUT RAINWATER HARVESTING AND ENERGY CONSERVATION"

in our planning application, and they were fairly supportive when they realised that we wanted to build a sustainable house."

For some time, Martin and Lucy had hoped to live a more environmentally friendly lifestyle, and building Pebblebed House offered the perfect opportunity to put their ideas into practice. The design was, therefore, very much eco-led, but was also influenced by the couple's relatively modest £200,000 budget.

"Building an eco house is definitely getting easier to do, but most builders still tend to suck their teeth when you start talking about rainwater harvesting and energy conservation," Lucy explains. "We undertook quite a lot of research and visited the Centre for Alternative Technology in Wales to find out what would be possible and what we could afford."

Lucy and Martin chose Paul Humphries Architects, a local practice known for its energy-efficient and sustainable designs, who worked with them to determine a brief for their new home. The footprint of the house is virtually the same size as that of the previous bungalow, but an insulated curved roof formed over glulam beams has enabled two bedrooms and a bathroom to be situated on the first floor while meeting planning requirements to remain below the chimney height of the original pitched-roof bungalow.

Reorientating the building ensures that it benefits from passive solar gain, with the main living room and kitchen positioned facing south-west and the smaller utility, wetroom and study located to the east side of the house. The curved living room is a single-storey element with a wall of full-height glazing, rooflights and a large overhanging flat roof of sedum matting – which encourages birds and butterflies, absorbs around 75% of water run-off and helps to keep the room at an even temperature. In addition, the couple have also covered a section of the larger curved roof with solar panels to heat their hot-water supply.

In order to save money during the build, they put their belongings into storage and moved into a caravan on site, which proved to be fairly grim. They then set about demolishing the old bungalow themselves, and managed to recycle many of the materials including timber, hardcore and tiles, before employing a local farmer-turned-builder who had been recommended by a friend for the complicated groundworks.

Despite the fact that he had never previously built a new house, Martin and Lucy were so impressed by the accuracy of his work that they asked if he would go on to complete the main structure of thin-joint solar blockwork. Compact thermo-reflective insulation was later fitted between the single skin of 215mm blocks and the internal plasterboard. With so many strange angles to accommodate, the project was not particularly straightforward, but the builder proved to be the perfect choice and was happy to take on any challenge without exceeding the fixed price he had quoted.

Many local village houses in this area have characteristic 'poppled' walls, where old river pebbles have been set into lime mortar to create an unusual and hardwearing finish, and Martin and Lucy were keen to follow this tradition. They used pebbles dug from their site and incorporated poppled battered walls and buttresses on the ground floor of the house, offset by areas of lime-washed render and horizontal larch cladding. Instead of gutters, the louvred metal rain-handlers send rainwater out in a fine spray.

"Martin had taken time out from work to project manage the build and planned to complete some of the physical work himself, including the ➤

Natural paints, limewash and hard-wax oils were used extensively, along with timber, slate and Marmoleum flooring over underfloor heating.

plumbing and dry-lining," says Lucy. "However, in February 2004, he suffered a major stroke which was a total shock and has left him wheelchair-bound. Not surprisingly, everything came to a sudden and complete halt on site. The scaffolding was arriving that day as the roof had been due to go on, and we were extremely fortunate that my brother and some very good friends stepped in after a few weeks and started things going again. I was at the hospital each day and wasn't able to offer much help at this point, but luckily Martin had kept excellent diaries and notes on the build, which were invaluable, and our builder was extremely helpful and sympathetic."

Fortuitously, Martin had been keen to make the ground floor of the house fully wheelchair accessible, following a brief time spent in a wheelchair with a broken leg, and incorporated extra-wide doorways and flat thresholds into the original design. By the time he was well enough to leave hospital, 17 months after his stroke, the majority of the ground floor rooms were completed and the former study had been converted into a third bedroom. Martin was absolutely delighted to see the progress which had been made on the house and, although he has yet to see the two first floor bedrooms with their curved ceilings, plans are afoot to eventually fit a stair-lift.

"IN THE FUTURE, IT WOULD BE NICE TO HAVE A SMALL WIND TURBINE TO PRODUCE ELECTRICITY"

**Left: A pond beneath the curved glass wall of the living room reflects light back into the interior. Top: A woodburning stove provides extra heat when required.**

Despite this unexpected turn of events, the couple were determined that the eco credentials of their home would not be compromised. Local craftsmen undertook the plastering, joinery and electrics, and Lucy and their friends completed the project using environmentally friendly paint, limewash and hard-wax oils.

A geothermal heat pump serves the underfloor heating and hot water, harnessing natural warmth from below the ground and supplementing the solar panels in winter. Even during the coldest weather, the temperature within the building remains consistent thanks to the combination of double and triple-glazed windows, a heat-recovery and ventilation system, insulation, passive solar gain and the sedum roof – a far cry from the couple's previous draughty Victorian house.

Outside, this effect is further aided by the new earth banks created to the

north of the house and built up in layers using soil excavated from the foundations. The site is relatively flat, open and exposed, and forming these sedum-covered banks to either side of the back door provides additional insulation and also helps to keep the larder cool – ensuring a degree of shelter to the most exposed wall of the building whilst helping it to recede into the natural landscape.

Two new ponds have also been dug in the garden. One is positioned outside the curved glass wall of the living room and collects rainwater from the sedum roof, reflecting light back into the building and acting as a cooling element in summer. The other was designed purely to encourage wildlife to the site and has already attracted a variety of newts, frogs and dragonflies.

"We put up bat boxes and bird nesting boxes," says Lucy, "and there are three composting areas which help us to grow organic vegetables such as onions, leeks, spinach and courgettes.

"We also have our own water supply thanks to an existing well, so we do feel as though we are relatively self-sufficient. In the future, it would be nice to have a small wind turbine to produce electricity, but for now we are more than happy with the house and our new way of living. In spite of what has happened, we still feel very fortunate indeed to be here and to be able to enjoy such a beautiful setting."

## FACT FILE

**Names:** Lucy Rutherford and Martin Hutchins
**Professions:** Film sound editor/carer and consulting engineer
**Area:** East Devon
**House type:** Three-bedroom eco house
**House size:** 155m²
**Build route:** Builder, subcontractors, friends and selves
**Finance:** Private and Norwich & Peterborough self-build mortgage
**Construction:** Lightweight insulating blockwork clad in pebbles, render and larch
**Warranty:** Architect's Certificates

**Build time:** Sept '03-July '05
**Land cost:** £190,000
**Build cost:** £200,000
**Total cost:** £390,000
**Current value:** £500,000
**Cost/m²:** £1,290

**22%**
**COST SAVING**

## FLOORPLAN

**The main living room of the house has been orientated south-west to maximise passive solar gain, with the neighbouring kitchen facing west. Behind these rooms are a utility, wetroom and bedroom, with a further two bedrooms and a bathroom on the first floor.**

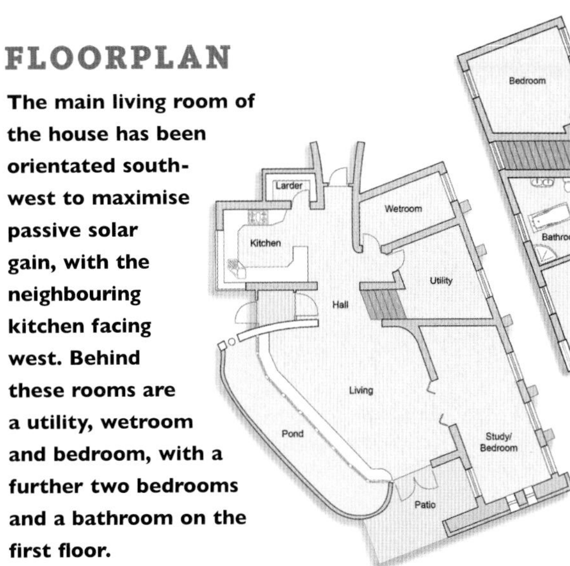


## USEFUL CONTACTS

Centre for Alternative Technology: 01654 705981 **Architect** Paul Humphries Architects: 01395 276598 **Builder** David Lupton: 01626 862191 **Roofs** Leicester Felt Roofing: 01823 662495 **Ground-source heat pump** Eco Heat Pumps: 0114 296 2227 **Windows** Swedhouse: 01905 791090 **Insulation** Actis Insulation Ltd: 01249 446123 **Solar blocks** H+H Celcon: 01732 886333 **Paints and finishes** Green Building Store: 01484 854898 **Underfloor heating** Nu-Heat: 01404 549770 **Solar panels** Construction Resources Ltd: 020 7450 2211 **Slate** The Delabole Slate Company: 01840 212242 **Louvred gutters** Rainhandler Europe Ltd: 028 9147 1212 **Heat recovery and ventilation** Villavent: 01993 778481 **Flat-roof system** Bauder Ltd: 01473 257671 **Kitchen** Alno: 020 8898 4781 **Beech staircase** Simon Kohn Furniture: 0117 987 3202.

# LAKESIDE LIVING

**Steve and Phillippa Lambert raced against the clock to rescue a historic garden and lake as a magnificent setting for their Japanese-inspired eco house.**

WORDS: DEBBIE JEFFERY
PHOTOGRAPHY: JESSICA DOBBS

When Phillippa Lambert first saw the St. Lawrence 'Tropical Bird Park' on the Isle of Wight, it had been closed to the public for seven years and had quickly become overgrown. "The mild climate had transformed the site into a romantic jungle, and the woodland walks and walled garden were full of disused bird cages," she recalls. "The lake was silted up, the fountain was no longer working and a huge black poplar tree had fallen into the water. It was like discovering a miniature Lost Garden of Heligan – and was very exciting."

As a landscape designer, finding the remains of this Victorian garden was a dream come true for Phillippa. She and her husband, Steve, a graphic designer, had been looking for a suitable site on the island where they could embark on a self-build project to create a new home and workplace, and Phillippa could immediately see the potential for restoring the significant gardens, which stand in a magical seaside setting within an Area of Outstanding Natural Beauty.

With no existing dwelling on the site, the couple needed to convince the planners to allow them to build a new house, and Phillippa spent several months researching the history of the garden before producing a definitive report, backed up by a number of landscape restoration drawings and a land-management plan. The planners were extremely supportive of the idea to give the site a new lease of life, and granted planning permission with the

The site is a two-acre lake and walled garden complex, part of the original designed landscape of a significant Victorian marine residence on the Isle of Wight.

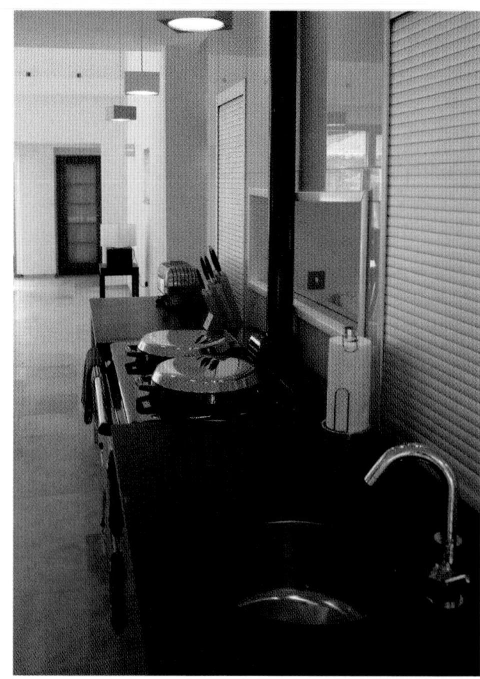

condition that the full landscape restoration should be completed within the space of just two years.

"We needed to work very quickly and decisively, and engaged Michael Rainey as our architect to design a suitable house," says Steve. "Our brief was to create a building full of light and space which would provide both living and studio areas. It should be cool and contemporary, with Japanese influences and – most importantly – blend seamlessly into the landscape."

Mike Rainey's brilliant concept sketch of an Oriental-influenced 'boat house', decking out over the lake to the west and overlooking the walled garden to the east, perfectly captured the spirit of the site and delighted everyone who saw it. Once planning permission had been achieved, the Lamberts turned to well-known eco architect Simon Clark, of Constructive Individuals in London, to produce the detailed specification and incorporate green methods and natural materials.

Work to restore the landscape to its former splendour began in earnest, and proved to be a monumental task to complete within the two-year deadline. Removing the huge fallen tree, catching and relocating the existing carp and dredging the many tons of accumulated silt out of the lake was a top

"WE HOPED THAT THE USE OF CLEAR AND FROSTED GLASS WOULD RELATE THE INTERIORS TO THE WATER OF THE LAKE OUTSIDE"

**Square suspended lights illuminate the kitchen work surfaces, and a reconditioned Aga has been installed against the free-standing wall between the kitchen and studio.**

priority, and months of painstaking work followed to clear the undergrowth and faithfully recreate the ornamental lake and its surroundings as Victorian owner William Spindler would have known it.

The two-acre site benefits from a magnificent natural landform of dramatic cliffs, which produces a protected microclimate where subtropical plants flourish within sight of the sea. Phillippa – who has been a garden and landscape designer on the island since 1986 – designed planting for the newly restored lake sides, as well as transforming the walled garden into a dazzling potager full of flowers, fruits and vegetables.

A new glasshouse graces the west-facing wall to complete the traditional look, and the site – which now also features a Japanese courtyard – acts as a showcase for Phillippa's design work and is also open on demand and without charge for schools, colleges, interested groups and charity fundraising.

For the first year of work, the Lamberts rented a property while they renovated an existing building on the site. "We had sold our old cottage and planned to live in the former bird park gift shop, but then there was a fire in the building two days before we were due to move in, so we had to find somewhere to rent while we worked on rebuilding it," remembers Steve.

This element now forms an attached two bedroom guest annexe where Phillippa and Steve lived for the remainder of the build. The couple decided to co-ordinate the various specialist subcontractors themselves, in addition to undertaking a tremendous amount of the labouring work, with the aim of creating a high-quality new build on a low budget.

"The greatest thrill was meeting and working with all the wonderfully

**Natural materials such as English limestone, untreated wood cladding and decking, slate, leather and oak flooring tie the interior to the surrounding landscape.**

The suspended woodburning stove creates an eye-catching feature in the living room, where black MDF provides stylish lids for the storage window seats.

**The huge master bedroom enjoys views to the west along the length of the lake, with full-height grey MDF wardrobes providing useful storage.**

"THE GREATEST THRILL WAS MEETING AND WORKING WITH ALL THE WONDERFULLY SKILLED CRAFTSMEN WHO HELPED MAKE THE BUILD HAPPEN"

skilled craftsmen who helped make the build happen," says Phillippa, who baked cakes for the builders throughout the project and was determined to avoid any disagreements on site. "The groundworks and foundations were monumental, and the steel element of the frame was faultlessly executed. The whole process was exciting, exhausting and unforgettable."

The highly insulated post and beam structure was erected on a reinforced concrete raft foundation, and has non-load-bearing infill walls and a sloping roof system. Steel beams jut out over the lake to carry the timber deck, which may be accessed through sliding doors in the sitting room/kitchen – creating an extension of the main living space which is perfect for eating and relaxing outside during warm weather as the water laps below.

March 2004 saw the completion of the steel frame, allowing the timber frame specialists to make rapid progress with installing the floor joists. By the end of the month the first roof joists had been slotted into position and the exterior walls were covered with breathable membrane and finished in a combination of mineral render and untreated horizontal timber cladding, fixed with stainless steel nails. With the main roof structure in position, the shape of the building could finally be appreciated – and the mellow Chinese slates which cover the roof are already blending in well with the surroundings. A central vacuum system was installed before the walls were closed for dry-lining, and a self-levelling screed covers the zoned underfloor heating which was laid throughout the house.

Glazing accounts for 36% of the overall floor area, the majority orientated west, and the interiors are filled with light and the reflections of the water – enhanced by a palette of natural materials and reflective glass surfaces. English limestone, untreated wood weatherboarding and decking, slate, soft paint colours and limed oak flooring are the key materials which tie the contemporary building to the surrounding landscape.

A great deal of effort has gone into making the structure as eco-friendly as possible, and the building employs renewable natural resources, including a suspended woodburning stove – which forms the focal point to the northern end of the main living space – organic paints and rainwater harvested for flushing toilets and irrigating the restored walled garden.

"We always knew that the house would have to earn its keep," says Phillippa, who designed the neutral interiors partly with the intention of offering the property as a photo shoot location. "Steve and I still both work from home but it's a very different lifestyle to the one we had before, and extremely busy and sociable. I did, however, design the gardens to be as low-maintenance as possible, which means that we do occasionally have time to simply relax and enjoy living in such a fabulous setting." ●

A new glasshouse graces the west-facing wall to complete the traditional look, and the site – which now also features a Japanese courtyard – acts as a showcase for Phillippa's design work and is also open on demand and without charge for schools, colleges, interested groups and charity fundraising.

## FACT FILE

**Names:** Phillippa and Steve Lambert
**Professions:** Landscape designer and graphic designer
**Area:** Isle of Wight
**House type:** Three bedroom-house
**House size:** 345m²
**Build route:** Self-managed subcontractors
**Finance:** Private plus materials sponsorship
**Construction:** Timber and steel frame, slate roof
**Build time:** June '03-July '05
**Land cost:** £125,000
**Build cost:** £370,000
**Total cost:** £495,000
**Current value:** £1,500,000
**Cost/m²:** £1,072

**67%**
**COST SAVING**

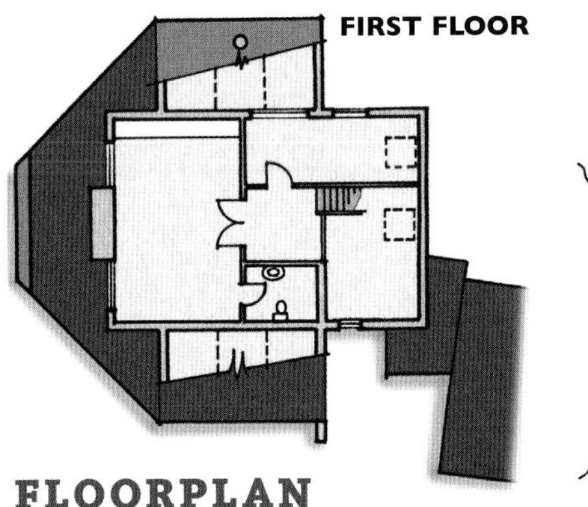


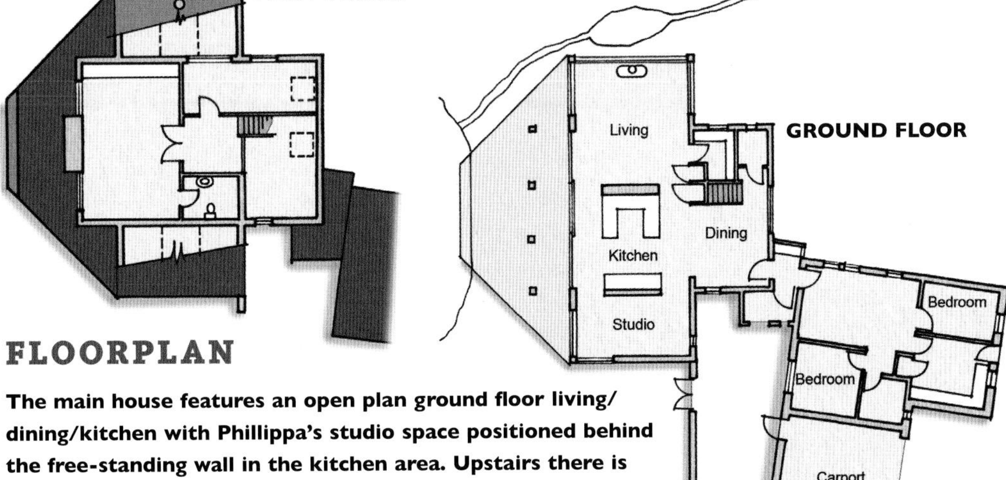


## FLOORPLAN

**The main house features an open plan ground floor living/dining/kitchen with Phillippa's studio space positioned behind the free-standing wall in the kitchen area. Upstairs there is one spacious bedroom, an en suite bathroom and Steve's studio, while a two-bedroom annexe provides additional guest accommodation linked to the main house.**

## USEFUL CONTACTS

**Lake House Design:** 01983 855151 **Concept architect** Rainey Petrie Johns Ltd: 01983 822882 **Project architect** Constructive Individuals: 020 7515 9299 **Structural engineer** Michael Long: 01983 840782 **Timber frame** Mathew Pitts: 07771 521462 **Structural steel work** Island Structural: 01983 525070 **Windows** Swedhouse: 01905 791090 **Underfloor heating** Thermafloor: 01738 620400 **Bathroom fittings** Vitra (UK) Ltd: 01235 750990 **Lighting design** Light Partners: 01983 855468 **Limestone flooring** Albion Stone Quarries Ltd: 01737 771772 **Organic paints** Auro Ltd: 01452 772020 **Insulation** Celotex Ltd: 01473 820820 **Mobirolo stairs** TB Davies (Cardiff) Ltd: 02920 713000 **Focus woodburning stove** Diligence International Ltd: 01364 654716 **Melamine-faced window seat cupboards** Egger UK Ltd: 01434 602191 **Custom-made external doors** Fairoak Timber Products Ltd: 01722 716779 **Prints on canvas to order** The Digital Room: 01455 828805 **Automatic irrigation** H2O Irrigation: 01788 510529 **Val-Eur furniture and accessories** InGarden: 01732 832299 **Composting and shredding equipment** Henchman: 01635 299847 **Thermowood cladding and decking** Finnforest UK Ltd: 020 8420 0777 **Wood flooring** Kahrs UK Ltd: 01243 778747 **Galvanised rainwater goods** Lindab Ltd: 0121 585 2780 **External render** Marmorit UK: 0117 982 1042 **Roof windows** Metal Window Company: 01993 830613 **Fitted kitchen furniture** Nolte Kuchen: 01279 868500 **Home ventilation system** Nuaire Home Ventilation: 029 2088 5911 **Automatic lighting control** Rako Controls: 0870 043 3905 **Integrated vacuum cleaner** Smart Central Vacuums Ltd: 01432 880090 **Glasshouse** Alitex: 01730 826900

# ECOLOGICALLY ORIENTATED

Greig Munro has designed a new home for his parents that combines sustainable building principles with a stylish exterior appearance: all for just under £163,000.

WORDS: CAROLINE EDNIE PHOTOGRAPHY: ANDREW LEE

Don and Ena Munro's new home – part of the Findhorn Community on the north-east coast of Scotland – was designed by the retired couple's son Greig and daughter-in-law Kathleen, otherwise known as the Findhorn-based architects Affordable TM. As Greig explains, Affordable TM certainly had their work cut out when it came to designing a house to fit within the specific and rather tricky Findhorn site which had come into his parents' possession. "If I was an architect designing a new house on a different location then I probably wouldn't have designed it quite like this, but we specifically designed it to fit within the site," explains Greig.

"On one side of the house we have a two-storey size house, on the other a small grass-roofed one-and-a-half-storey house; we designed our house to make a meaningful join and connection.

"A lot of the houses in the Findhorn Foundation look quite Scandinavian or North American in terms of their style and in the materials used and their roof pitches," Greig continues. "With Don and Ena's house we wanted to maintain a balance, but also to create something that was intrinsically Scottish in character. So, I think this is reflected in the gables and steeper roof pitch, as well ➤

Large areas of glazing to the south help the light and heat from the sun flow into an atrium space, enabling the benefits to be felt throughout the house.

"THE SOUTH-FACING FRONT BOASTS A COMPACT HORIZONTALLY CLAD WING THAT CULMINATES IN A 'SUN TRAP' DECK"

as the proportion of the windows – there is not too much glazing."

As a result, the Munro house is a neatly bespoke two-storey, two-bedroom home with a 200mm insulated timber stud frame. In addition to the steep plain clay tiled roof, the house is characterised by vertical and horizontal cladding made of locally grown Scottish Larch. The south-facing front boasts a compact horizontally clad wing that culminates in a 'sun trap' deck, while a vertically clad gable succeeds in creating a real sense of height. The Velux glazing, very generous on the south-facing front, is finished via copper flashing, as are the patinated rainwater goods. These copper hues blend harmoniously with the blue grey tint of the weathering larch and the blue-painted balcony and decking canopy.

In terms of the interiors, Greig admits that his parents had a clear idea about how they wanted to live in their first self-build. "They've owned lots of houses – 20 in all – the best aspects of which we've tried to jigsaw together into this one house. Our job was to take these bits of ideas and put them together," Greig explains. "My parents have never lived in an open plan house, built a new house from scratch or taken on the idea of an eco house, so it was quite an adventure for them."

This first self-build 'adventure' for Don and Ena involved downsizing from their previous four-bedroom house to a two-bedroom arrangement. In addition, they were keen to create a relationship between the lounge and living space on the ground floor, and to have a large 'social' kitchen. The result is a ground floor which contains the desired roomy open plan, south-facing kitchen which leads to a ground floor lounge area which expands seamlessly out to a sizeable deck. A blockwork core element contains the chimney from the ground floor stove. Its location at the centre of the house creates an effective radiating heating system from the centre to the periphery, and in addition there are no doors in this area in order to allow the air to circulate. "I hate corridors anyway!" admits Greig.

Completing the picture on the ground floor is a bedroom with a bathroom that can be accessed from the bedroom or hall. A porch has been created in the north entrance to the house and as Greig explains,

**A central core chimney has been designed to enable heat from the stove to radiate throughout the rest of the home.**

**A large 'social' kitchen leads to a ground floor lounge area which expands seamlessly out to a sizeable deck.**

"DURING THE DAYTIME WE SPEND OUR TIME DOWN HERE AND IN THE SUMMER WE SPEND A LOT OF TIME ON THE SOUTH-FACING DECK HAVING COFFEE OR LUNCH"

"There's so much sand around here from the beaches so nearly all the houses have a 'shoes-off' place. The porch is like a room in itself, and because people drop in quite a lot, if you don't want them to come into your home, this area is big enough and light enough to have a chat, without necessarily having to invite someone into the main area of the house. This means you can control your own space."

A staircase with hand-crafted balustrade in ash made by local artisan, Nigel Hilton, leads to the first floor from the hall. This upper level principally features an impressive main living area with south-facing sea views, and en suite master bedroom. "My mother is small and she's always wanted a house that she felt was designed around her. In the master bedroom there's a low level window and this means they can both sit up in bed and look out to the deer in the field, and over out to the Findhorn Bay," says Greig.

Throughout the house the colour palette and finishes have been kept simple and natural. The hard-wearing elm floors provided by local sawmill Norbuild FTC have been crafted from a single wind blown tree. Indeed, Greig admits that while designing the house, "It was about looking for what's available locally and making sure we could accommodate it. For example the timber members are C16 grade which is the only grade of structural timber that you can get in Scotland, so we made sure we could source the timber locally."

Although the main lounge is situated on the first floor, Don and Ena have found themselves gravitating towards the small ground floor living space. "The kitchen and ground floor lounge has become our main focus," explains Don. "During the daytime we spend our time down here and in the summer we spent a lot of time on the south-facing deck having coffee or lunch. It's one of the driest and sunniest areas of Scotland and since it's so easy to move in and out of the ground floor lounge to the deck, it's ideal. It works really well."

Another aspect of the house that works really well is its ecological and energy efficient performance. The Findhorn Ecovillage has been something of a pioneer in terms of championing ecological building solutions – 60 eco-homes have been built to date – and the Munro house is no exception. In terms of Don and Ena's eco credentials, solar heat collector panels are located in the roof and a solar calorifier tank – which acts like a heat ➤

**Although Don and Ena tend to use the ground floor living spaces more frequently, the more formal first floor living room enjoys pleasant views and opens out onto a balcony.**

"ANY ELECTRICITY REQUIRED TO TOP UP THE SYSTEM COMES FROM THE COMMUNITY'S OWN WIND TURBINE"

exchanger – is situated in the ground floor utility room. The hot water is heated up by the solar panels on the roof and this runs the thermostatically controlled central heating system – any electricity required to top up the system comes from the community's own wind turbine. For an extra boost in the winter there's the wood burning stove in the ground floor lounge and a small gas fire upstairs in the main living area.

The house is also heavily insulated by 200mm of sheep's wool. "It's lovely stuff," explains Greig. "This is the first house in Findhorn that uses Second Nature Thermofleece. In fact, I think it's the first one in Scotland. I chose this type of insulation over others such as recycled paper because I don't believe paper passes moisture in the same way. When I go walking in the hills I don't wear a newspaper, I wear a woolly jumper, and wool gets warmer as it gets moister. So I'd been looking for a product like this and I believe it works. In terms of providing a breathing wall construction it's excellent and robust. The joiners love it too as it's non-irritant as well as being lovely to touch. This is a very well insulated timber frame house."

Finally, in terms of the building's exemplary eco credentials, all interior paints are low toxic silicate based and these act as an internal vapour check to prevent moisture going into the fabric too quickly.

In terms of the construction process, the building took six months to complete – the water, waste and electrics were all provided in the land cost. And in addition to seeking local authority planning approval, the Munros also had to submit their designs to an internal Findhorn Park planning group. Their role is to essentially check any new build project's eco credentials and solar proximity layout in relation to neighbouring houses.

Indeed, the whole self-build experience seems to have been a very enjoyable one for all involved. Don, who decided to opt for this lifestyle change by commissioning the new build on the year of his 70th birthday, seems to have particularly enjoyed being involved in many aspects of the design and supervision of the construction.

"I wasn't the supervisor but maybe I was advisor to the supervisor," he says. In fact, although Don concedes that he and Ena are extremely enamoured of their community and their new eco-home, when asked if this is his first and last self-build, he cracks a 'never say never' smile.

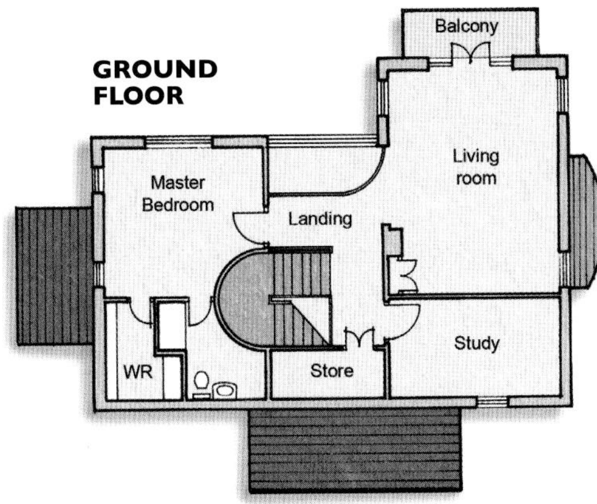


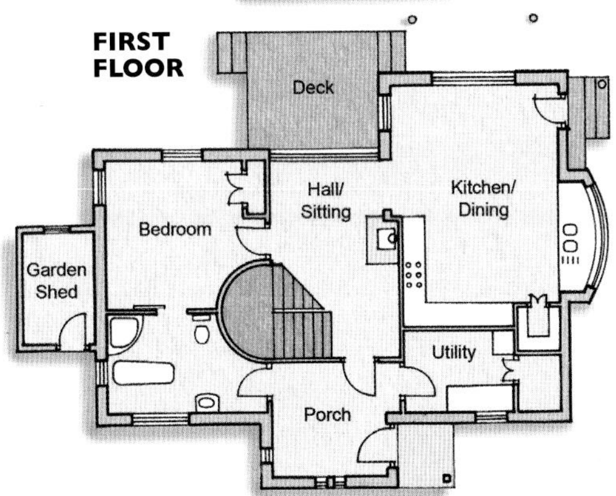


## FLOORPLAN

**The ground floor includes a self-contained bedroom suite which would meet the owners' future requirements should the need arise. The upstairs living room enjoys wonderful views and opens out onto a balcony.**

**Architect Greig designed the house for his father Don (left) and mother as part of the eco-community at Findhorn in north-east Scotland. The home contains a raft of features designed to save energy and promote sustainable development.**

## FACT FILE

**Names:** Ena and Don Munro
**Professions:** Retired
**Area:** Morayshire, north-east Scotland
**House type:** Detached
**House size:** 147m²
**Build route:** Main contractor with local suppliers
**Construction:** 200mm thick insulated timber frame with timber cladding
**Finance:** Private
**Build time:** Jan '04-July '04
**Land cost:** £38,000
**Build cost:** £162,560
**Total cost:** £200,560
**House value:** £220,000
**Cost/m²:** £980

**9%**
**COST SAVING**

**Cost breakdown:**

| | |
|---|---|
| Enabling work | £3,500 |
| Substructure | £10,700 |
| Frame and external walls and joinery | £56,000 |
| Windows, blinds, doors | £17,000 |
| Roofs | £14,000 |
| Wall, ceiling, floor finishes | £24,000 |
| Sanitaryware, kitchen, built-in fittings | £14,360 |
| M&E, plumbing, electrics | £19,000 |
| Landscaping and external works | £4,000 |
| **TOTAL** | **£162,560** |

## FINDHORN FOUNDATION

The Findhorn Community was set up in 1962 to provide an alternative holistic and sustainable model for living for its founders. These days the Findhorn Foundation, established by the community in 1972, is one of the largest holistic communities in the world. It also boasts one of the UK's most progressive eco-villages which provides a focal point for some of the cutting-edge developments in the sustainable building sector. www.findhorn.org

## USEFUL CONTACTS

**Architects** Affordable TM Architects Ltd: 01309 692240; **Structural engineer** AF Cruden Associates: 01463 719200; **Main contractor** John Duncan Construction: 01542 833501; **Exterior larch cladding, elm stair and hardwood flooring** NorBuild: 01309 671700; **Roofing** – 20/20 Plain Clay Tiles by Sandtoft Roof Tiles: 01427 871200; **Windows and external doors** NorDan: 01224 633174; **Heating** Solar collectors from AES Solar: 01309 676911; **Solar calorifier tank** Rotex 'Solaris Sanicube' by Thermalfloor UFH Systems: 0845 062 0400/01738 620400; **Kitchen** Riverside Kitchens: 01343 552202; **Sheep's wool insulation** Second Nature UK: 01768 486285; **External vapour barriers** Klober: 01509 500671; **Stair handrails, balustrades and timber support brackets** Nigel Hilton: 01309 671442; **Wood stove** Bonk & Co Limited: 01463 233968

Nicholas Worsley has realised his dream of building an eco-friendly home, using sustainable natural materials.

# DOWN TO EARTH

WORDS: CLIVE FEWINS PHOTOGRAPHY: JEREMY PHILLIPS

Is this the riding stables?" is the most commonly asked question when lost visitors stumble up an unsurfaced lane and stop outside the unusual-looking new house of Nicholas Worsley. He usually redirects them with the words: "No, but you couldn't have said a nicer thing!"

Nicholas, a retired barrister, loves occasions like this. They are the very essence of why he built his new house, on a low bluff above the banks of the River Severn near Worcester. The house is bounded by a tall, 70m-long boundary wall of cob – a mixture of earth, straw and gravel – which also forms some of the inside wall of his wine store and utility room and also one of the guest bedrooms.

"Humour, mystery, fantasy, ecological, sustainable and as far as possible, invisible." This was the 11-word brief Nicholas gave to his architect and long- ➤

"HUMOUR, MYSTERY, FANTASY, ECOLOGICAL, SUSTAINABLE AND AS FAR AS POSSIBLE, INVISIBLE"

**The house, constructed mainly of timber frame, also incorporates a 70m-long boundary wall of cob.**

standing friend John Christophers at their first design meeting after he had found the plot in 1999.

**Concrete kitchen worktops were chosen for their effect and qualities as a thermal store; the concrete island unit was cast in situ.**

He and his late wife, Anna, had been scouring the outskirts of Worcester for three years after deciding to quit their three-storey Georgian townhouse in the city in which they had lived for 31 years, as she was wheelchair-bound and unable to get about.

When Nicholas found the site – it measures just under a hectare – the owner, a local farmer, had unsuccessfully tried to sell it for development, and Nicholas enlisted the help of officials and Worcester City Council to indicate what kind of development might be acceptable.

"John Christophers did most of the negotiation but it was not easy as there was a presumption against building on the site," Nicholas says. "The key thing was that it had to be the right design."

Fortunately what the city officials favoured coincided almost exactly with what Nicholas, who has a lifelong interest in art and architecture, had in mind. "It had to be on one level to accommodate my wife's wheelchair, a rather angular shape, to suit me and to contrast with the softness of the cob, and a sort of succession of house/garden outdoor rooms as you move around," he explains. In addition, he wanted it to have as many ecological features as possible.

At the same time it was not specifically designed to be a courtyard house. Its U-shaped plan and curved cob wall, plus the sharp angles, put paid to the

"THERE WAS A PRESUMPTION AGAINST BUILDING ON THE SITE. IT HAD TO BE THE RIGHT DESIGN"

"IT HAS FULFILLED ITS OBJECTIVES. IT IS RELAXED, UNDEMONSTRATIVE, ECO FRIENDLY AND NOT SHOWY"

regular-shaped spaces demanded by a series of courtyards.

Drive along the unmade approach road that skirts comfortable detached homes in this select outer corner of the city and you could well mistake the house, with its rough cob-encircling wall and ribbed aluminium capping, for a series of converted agricultural buildings – or, indeed, a riding stables.

Inside all is dramatically different – but first you have to find the front door. It is, as Nicholas describes, "undemonstrative". You have to peer through what looks like a very ordinary farm building entrance. A square-headed arch – very understated – leads to an inner paved area, where an unobtrusive curved timber door beckons. Inside you find yourself in a circular top-lit vestibule beneath a domed rooflight with three exits. Once you have gained entry, returning to the vestibule after a few minutes to try to find which door leads to the outside world is an interesting experience.

**The flat roof is punctuated by a series of rooflights and lightpipes. "The amount of vertical light we get in this house is quite phenomenal," says Nicholas.**

Inside there are all sorts of quirky touches, such as angled bookshelves in the study that Nicholas designed to take different sizes of book. Cast concrete kitchen worktops are a good example of the blend of design and ecological elements that both Nicholas and John aimed at. The material was chosen for its effect and thermal mass. The kidney-shaped island unit, which was cast and ground in situ, is a tour de force.

The flat, thick, membrane-covered and heavily insulated roof is punctuated by a series of rooflights – one tall enough to create a special space to accommodate a particularly tall picture in Nicholas' collection. There is also a four-metre-high walk-in shower room which is so tall that it towers above the rest of the roof.

It was designed for wheelchair access and to take a hoist for Anna. Sadly she died within a year of their move. Although this room is, therefore, tinged with sadness, it is still one of Nicholas' favourites. It is top lit with the aid of three lightpipes. "The amount of vertical light we get in this house is quite phenomenal," says Nicholas.

Turning to the 'eco' features of the house, the cob was chosen because it is a renewable material, it has good insulative qualities, and it is far cheaper to ➤

The cob wall is used as a feature in its own right in the guest bedroom.

**The internal vestibule, top lit by a light pipe, has three doors. Finding the right one to the living spaces is all part of the fun nature of the design.**

"A LOT OF SO-CALLED ECO HOUSES ARE VERY BORING ARCHITECTURALLY. I LIKE TO THINK THIS HOUSE IS THE REVERSE"

build than a masonry wall. The clay came from a local construction site. "I spotted it on the site of a new factory building in Malvern," Nicholas recalls. "It cost us nothing. We only had to pay for the gravel and the straw – and, of course, the Devon-based team that built it as there is no cob building round here."

The building is also designed to make maximum use of solar gain. An array of double-glazed windows faces towards the river and into the sun around the series of angles. To the other side, the deep cob wall keeps the north cold at bay and insulates the building. There is a thick layer of recycled newspaper insulation in the deep roof, which an array of evacuated tube solar panels on the kitchen roof ensures hot water all the year round whenever the sun is shining.

"It produces a pretty good flow of hot water and means that in summer and for much of the rest of the year I use virtually no space heating: the only gas I use is for cooking," says Nicholas. "In fact with the glazed southerly aspect the house is so hot in summer that I need the exterior vines on their overhanging wire structures to provide seasonal shading. I did, at first, keep a record of how much gas I was using. In the summer I used about one unit a week."

The gas-fired underfloor heating was installed primarily because of Anna, but since she died Nicholas has found he has used it very little – perhaps when entertaining in winter or in the visitors' rooms.

Apart from the cob exterior wall, the house is timber framed. "The seasoned oak for the cladding was sourced locally," says Nicholas. "I love the way it has been fixed horizontally using the 'coach screw' method. John drew out every detail of this and the rest of the house, which is probably why the chamfered edges of the oak weatherboard fit together so well. The colours it has produced after three years of weathering are splendid."

Thermal mass for the storage of interior heat is provided by the 750mm thick cob wall, the insulated concrete floors, concrete kitchen worktops and chimney breast. Rainwater from the roof is collected and used in the WCs, washing machine, and for garden irrigation.

A combination of all these factors won the house a 2003 City of Worcester Award. It was cited as an example of 'an outstanding work of modern architecture capable of becoming a building of wider significance'. It also was praised for its 'comprehensive commitment to the principles of sustainability'.

Architect John Christophers was delighted by this because his intention throughout had been to try and combine inspired design with a project considered sustainable. As Nicholas puts it: "A lot of so-called eco houses are very boring architecturally. I like to think this house is the reverse." He adds: "It was not cheap, but it is quite a big house, and we have a fifth en suite bedroom in the adjoining outbuilding where Anna's carer used to live. It has fulfilled all its objectives. It is relaxed, undemonstrative, eco friendly and not showy. It was the sort of site I never really dreamed of being able to obtain. You really do have to do something creative on a site like this."

## FACT FILE

**Name:** Nicholas Worsley
**Profession:** Retired barrister
**Area:** Worcester
**House type:** Five-bedroom detached
**House size:** 220m$^2$
**Build route:** Main contractor
**Construction:** Timber frame and cob
**Warranty:** Architect's Certificate
**Sap rating:** 100
**Finance:** Private plus sale of house
**Build time:** Jul '00-Sep '01
**Land cost:** £120,000
**Build cost:** £274,000
**Total cost:** £394,000
**House value:** £650,000
**Cost/m$^2$:** £1,245

## COB WALLS

Cob is a much-favoured material with self-builders looking to create an ecologically sustainable dwelling with high energy values: as a material it has higher U-values than alternatives such as rammed earth, as there are more air voids within its structure. Cob incorporates site soil and locally sourced clay, with straw acting as the binding ingredient. It is a wholly natural building product.

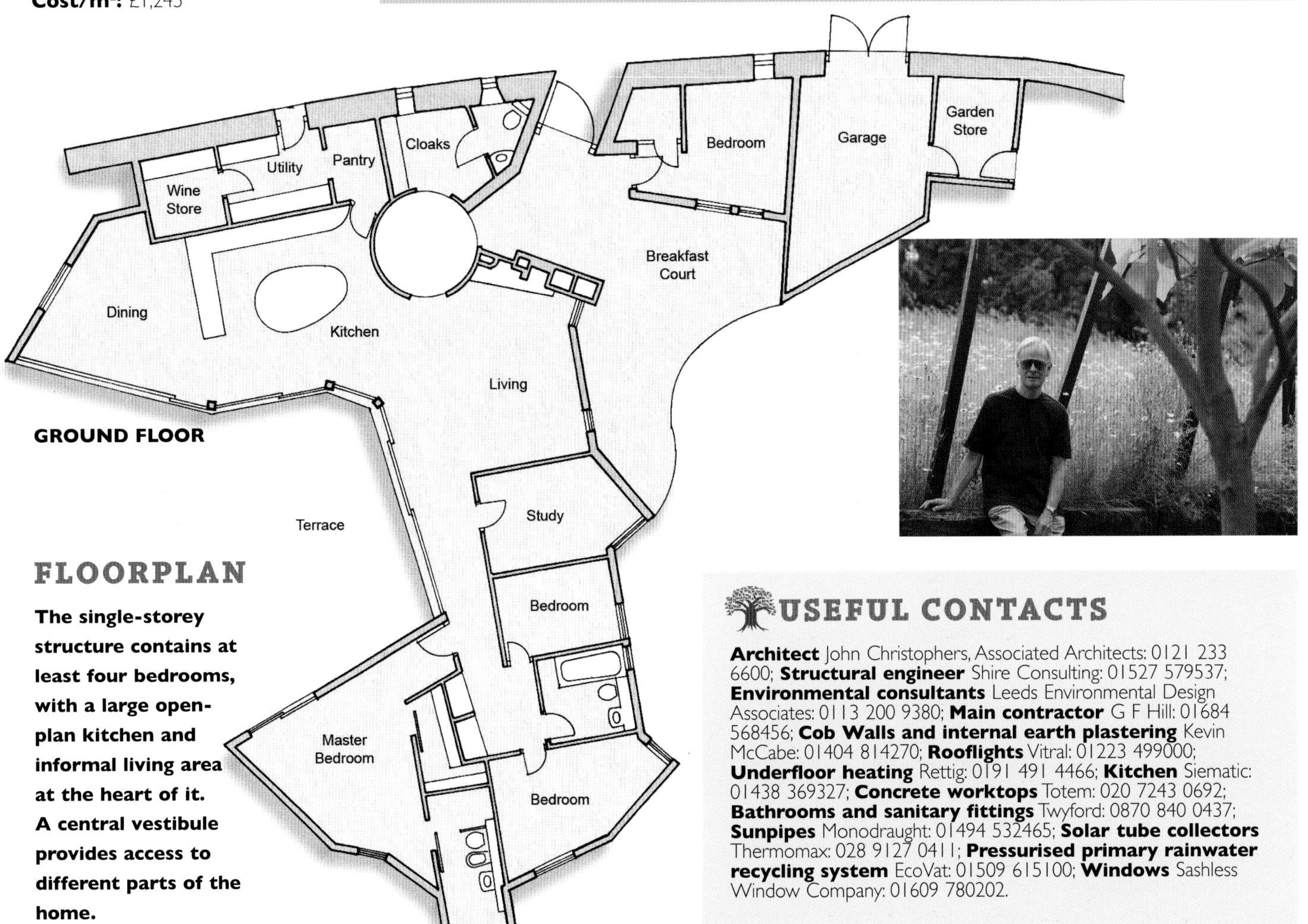


**GROUND FLOOR**

## FLOORPLAN

**The single-storey structure contains at least four bedrooms, with a large open-plan kitchen and informal living area at the heart of it. A central vestibule provides access to different parts of the home.**

## USEFUL CONTACTS

**Architect** John Christophers, Associated Architects: 0121 233 6600; **Structural engineer** Shire Consulting: 01527 579537; **Environmental consultants** Leeds Environmental Design Associates: 0113 200 9380; **Main contractor** G F Hill: 01684 568456; **Cob Walls and internal earth plastering** Kevin McCabe: 01404 814270; **Rooflights** Vitral: 01223 499000; **Underfloor heating** Rettig: 0191 491 4466; **Kitchen** Siematic: 01438 369327; **Concrete worktops** Totem: 020 7243 0692; **Bathrooms and sanitary fittings** Twyford: 0870 840 0437; **Sunpipes** Monodraught: 01494 532465; **Solar tube collectors** Thermomax: 028 9127 0411; **Pressurised primary rainwater recycling system** EcoVat: 01509 615100; **Windows** Sashless Window Company: 01609 780202.

# HOME GROWN

Tom and Sally Beevor have built a beautiful timber-frame home designed to have minimum impact on the environment and its parkland surroundings.

WORDS: DEBBIE JEFFERY
PHOTOGRAPHY: ROB JUDGES

**Larch from the Beevors' own land was used for the timber frame and cladding, while clay pantiles were reclaimed from an old barn on the family estate.**

"Building this house has been a liberating experience," Sally Beevor explains. "Our previous home, Hargham Hall, was a beautiful Queen Anne property. We loved living there, but it offered no opportunity to experiment with modern colours or to introduce energy saving ideas because the house dictated the style of decor. Here, we have been able to let our imaginations run wild!"

Hargham Hall in Norfolk has been home to the Beevor family since it was built in 1690, and is now occupied by their son, Hugh, his wife and three children. The couple had often thought that a derelict woodyard halfway up the drive to Hargham Hall would make a wonderful site for a new house, and decided to try to obtain planning permission for a dwelling. This was granted as a replacement for a dilapidated cottage on the estate, situated on the edge of a new motorway development. The cottage was demolished and the local authority agreed to allow its replacement to be sited in the more appropriate setting.

"Normally, we would only have been allowed to build within 500 metres of the existing building," says Tom, "but it was in such a poor location that the planners agreed to the idea. Without the cottage it's doubtful whether we would have been given permission to build at all, so we consider ourselves extremely lucky."

The Beevors knew that they wanted to build a wooden house which would fit in well with the surrounding landscape. Sally is a retired osteopath, and Norfolk-based architect Neil Winder was a patient who had purchased timber from the Beevors' woodyard. "Neil specialises in eco-friendly architecture, and we visited his house and fell in love with it," says Sally. "We just said, 'Please can we have one too?' Once we had seen his work we decided not to approach any other architects on our list. It was such a simple, comfortable timber house filled with colour and light."

Tom and Sally sat down with Neil and

"NOBODY WOULD EVER KNOW THAT THIS IS AN ECO HOUSE AS EVERYTHING IS EXTREMELY UNOBTRUSIVE"

**Sally was keen to exercise her creativity and, as a former kitchen designer, felt confident in designing the kitchen units, which were made by a local joiner and painted in a variety of colours.**

Colour is an important element of the interior design, with every room featuring several different shades from a range of paints inspired by the Swedish painter Carl Larsson.

**Three independent partitions in the living room have been painted in vibrant pink, soft blue and deep green and have small spyholes cut into them.**

## "ECO-FRIENDLY BUILDINGS OFTEN INVOLVE COPYING TRADITIONAL METHODS OF CONSTRUCTION USING GOOD QUALITY LOCAL RESOURCES AND SIMPLE TECHNIQUES"

designed The Old Woodyard, which is strongly based on Neil's own home. They requested a light, airy building which would relate well to its parkland setting and give easy visual access to the surrounding garden. "I have always been interested in the natural world, and Tom loved the idea of using locally grown resources to build something which would enhance his beloved Hargham," Sally explains. "The eco features were a natural progression. Once we had seen that Neil's house could be beautiful and ecologically friendly at the same time, it seemed an obvious choice. Apart from the solar panels on the roof nobody would ever know that this is an eco house as everything is extremely unobtrusive."

Environmentally friendly features include high levels of insulation made from recycled newspaper, non-toxic paints and stains, a solar hot water system linked to the central heating and a small, sophisticated, weather-linked oil-fired boiler. A reed bed sewage and compost system has been installed coupled with low-flush toilets, and all waste from the house passes through an Aquatron separator, which enables the flow of water to be diverted to the reed bed and the solids to be composted.

"There is nothing very radical about timber-framed houses," Neil explains. "Eco-friendly buildings often involve copying traditional methods of construction. I believe in using good quality local resources and simple techniques just as they did two or three hundred years ago. The Old Woodyard was put together by three men armed with hammers and a stack of nails, and most of the timber used in its construction was grown and processed here on the estate."

Two tumbledown timber barns on the site were demolished and the new house designed with single-storey wings flanking a two-storey central core. Clay pantiles were reclaimed from old farm buildings on the estate and the breathing wall structure clad in larch weatherboarding from the Beevor's woodyard and painted with a grey eco-friendly stain.

"Neil and I came up with the team of tradesmen between us," says Sally, who worked full-time project managing the build. "It was very convenient living at the Hall and being so close to our site, and it meant that we could take our time and ensure that everything was just as we wanted without feeling stressed by deadlines."

Instead of standard concrete foundations the larch timber frame is raised up on steelwork. Twelve posts sit on concrete pads and act like stilts, which improves the view out of the windows and reduces the impact of building on the site. Neil specialises in designing houses capable of withstanding the effects ➤

**The Beevors have incorporated a swimming pond – as opposed to a pool – which contains one specially treated section to separate it from the wildlife.**

"SWIMMING IN THE POND IS NOTHING LIKE A SWIMMING POOL, WHICH CAN BE BORING AND RATHER UGLY"

of climate change, such as extreme temperature fluctuation and – in addition to accommodating a large abutting ditch and a range of trees growing close by – the stilts ensure that the house will never be affected by flooding.

The house has been designed as a low-key affair, with no formal entrance hallway. A door opens directly into a lobby, which is used as a utility area, and ramps progressively lead up into the open plan main body of the house. "One of the key features is the use of independent partitions," says Tom. "None of the internal plasterboard partitions reach the ceiling or touch the external walls, which gives a sense of space while maintaining a degree of privacy. It's cosy, and yet open and light – which is a very clever combination." Tiny spyholes have been cut into these partitions to further increase the connection between spaces, and each partition is painted a different colour in contrast to the white of the remaining walls.

The finished house is bright and airy, providing a study for Tom and an office for Sally. Large windows connect it to the natural garden and wild flower lawns outside, which feature a large swimming pond. "Swimming in the pond is nothing like a swimming pool, which can be boring and rather ugly," Tom explains. "From the house you walk out over a ramp to the pool, which was designed by a specialist company to look like a beautiful, planted pond and is full of wildlife. One section is divided under the water and is kept clean by a pump filtering through gravel.

"We were slightly appalled at how little we regretted leaving our old home," says Tom. "Because this house is such a joy to live in and so easy to heat and maintain." ●

## FACT FILE

**Names:** Tom and Sally Beevor
**Professions:** Retired and partner in architectural practice
**Area:** Norfolk
**House type:** Three-bedroom detached
**House size:** 210m$^2$
**Build route:** Self managed subcontractors
**Construction:** Timber frame
**Warranty:** Architect's Certificate
**Finance:** Private
**Build time:** July '00-December '01
**Land cost:** Already owned, estimated value £150,000
**Build cost:** £250,000
**Total cost:** £400,000
**House value:** £500,000
**Cost/m$^2$:** £1,190

**20%**
**COST SAVING**

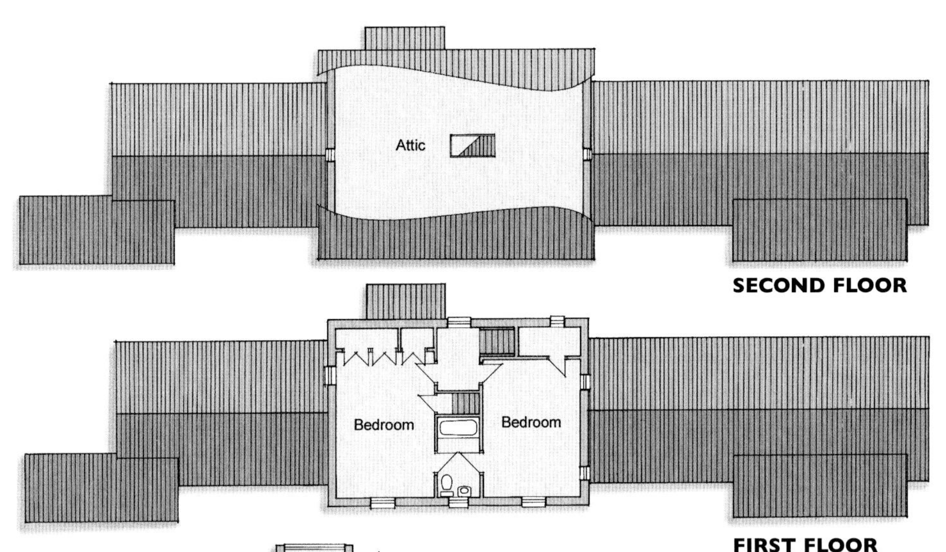


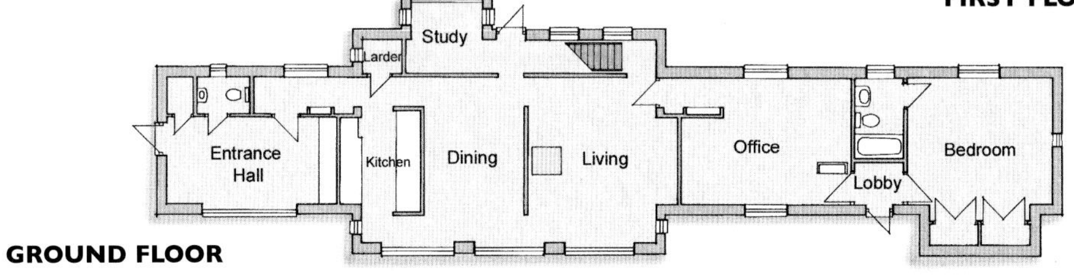


## FLOORPLAN

**The three-bedroom house has an open plan living space on the ground floor, with free-standing internal partitions offering privacy.**

## AWARD-WINNING DESIGN

This is a high quality project that manages to successfully combine all of the different elements on its owners' brief – elegant, traditional exteriors, informal, open-plan interiors and low environmental impact.

Timber frame construction was chosen because of its inherent ability to incorporate high levels of insulation and because of the availability of materials to hand from the surrounding woodland. As well as ease of assembly, lightweight construction also met the objective of minimal environmental impact – the house sits on minimal concrete padstones.

The Old Woodyard is a highly successful and environmentally responsible new house that ideally meets the needs of its owners. Despite its extremely sound ecological credentials, it manages not to compromise on aesthetic appeal, practicality or value for money.

## USEFUL CONTACTS

**Architect** Neil Winder, Greenyard Architecture: 01953 888386; **Structural engineer** A J Power & Associates: 01953 887539; **Groundworks** F G Butler: 01953 454168; **Builder** Kevin Blinks: 01953 454262; **Heating and plumbing** C J Bird: 01953 881040; **Electrics** J. Holdom: 01953 454247; **Specialised joinery** Stratton Period Joinery: 01953 860870, P. Brown: 01379 740590, G. Richmond Shaw: 01953 860110; **Sewage system** Watercourse Systems Ltd: 01984 629070; **Insulation** Payne Insulation: 01603 743407; **Garden design** Will Draper: 01953 789910; **Natural swimming pond** Metamorphosis: 01453 840351; **Metalwork** South Farm Engineering: 01953 717497

# FLYING THE DIY FLAG GREEN

**Adrian Thurley's new home – on which he worked almost full-time on a DIY basis – is packed with energy-saving features.**

WORDS: MARK BRINKLEY PHOTOGRAPHY: ROB JUDGES

Like many airline pilots, Adrian Thurley spent a lot of time up in the sky with his thoughts close to the ground, in his case dreaming of the house he would one day build. During the 1990s, the Thurleys were living in Bournemouth and were always on the lookout for building plots. However, Bournemouth is not an easy place to track down nice plots and Adrian wasn't having any success. His parents, meanwhile, continued to live in the same bungalow where Adrian had grown up – in a village near Newmarket, Cambridgeshire – and were beginning to find their 1950s bungalow and its large garden a little bit too much to handle and were contemplating selling up and moving to something smaller. Adrian realised it was his self-build moment. The plot his parents' bungalow sat on was easily large enough for the house he wanted to build, the location was manageable for Heathrow and Gatwick and, of course, he still had ties in the village.

The bungalow had been largely untouched over the years and, in planning terms, was an ideal candidate for replacement. However, it was only 100m² and the council, East Cambridgeshire, had a policy stating that replacement dwellings could only be 25% larger and not significantly taller. Negotiations went on for eight months in 2001 with Adrian and his designer, Roger Peyton, edging forward their plans for a much larger house. Adrian recalls: "The initial reaction of the planning officer to our plans was very negative: 'Far too big!' she exclaimed. So we moved the position of the house further back up the plot; this reduced the impact of the increased height. Then we agreed to step down the roof line a little and reduce the size of the front elevation by giving the house two rear additions, making it a U-shape. Finally, we went for a basement and a loft conversion to get the space we wanted."

However, the proposed house was still too big, and the planning officer had to refer the application to the committee. Adrian wrote explaining that the new house would incorporate a number of sustainable features that

**The new house sits on the former site of Adrian's parents' bungalow. Adrian and his designer had to engage in a protracted series of negotiations in order to build the new home, which exceeds the local council's standard replacement parameters.**

**Top: The fireplace, installed by Adrian and his team, was designed by the Fireplace Design Consultancy.**
**Below: the house is surrounded by French drains.**

should mean that it would use less energy and less water than the existing bungalow, even though it was four times the size. The planning committee paid a visit to the site in September 2001 and were impressed enough to grant permission later that day. It helped that the planning officer had recommended approval beforehand. All that the council asked in return was that the external appearance of the house should be largely traditional.

The Thurleys were fortunate in that the position of the new house was entirely behind the existing bungalow, which meant that they didn't have to demolish it until after completion. Even so, the bungalow wasn't big enough for everyone, so they started work by building the detached garage, which they then camped in for 18 months.

The building work was carried out by specialists from around the country together with a core group of local tradesmen and a huge input from the Thurleys. "I have, on average, worked on the house full-time three to four days a week (during the day only) for 120 weeks so far. Bev has also put in a considerable amount of time and my daughters have done their fair share!"

Adrian's project includes a fully submerged 75m$^2$ basement, housing a state-of-the-art heat pump and a rainwater harvesting system. The basement was built using ThermoneX pre-cast panels, craned into place on a raft foundation. "I think the basement has been tremendous value, even though I underestimated the amount of excavation. Its total cost is around £33,000 which makes it cheaper to build than the above-ground parts of the house."

The house is heated entirely by an Ice Energy heat pump. The heat pump is located in the basement; it's connected to two 200m long pipes that run under the front garden where the original bungalow once sat. It is capable of providing 18.5kW of heat using less then a third of this amount of electricity to do so. Adrian has been monitoring the fuel bills during their first winter in the house. "It's hard to be precise about how much electricity the heat pump is using because we are an all-electric house and we are running appliances and lights as well as the heat pump. But in the summer our weekly fuel bill was around £28 a week and during the coldest week this winter it rose to £65." This may sound like a lot but, bear in mind, this is a big house and the efficiency of the heat pump looks to be excellent with a space heating load of around 50kWh/m$^2$/annum compared with around 65kWh/m$^2$/annum which a gas condensing boiler would deliver.

Also housed in the basement are three rainwater harvesting tanks, supplied by Freewater UK. The cost for the whole system was around £2,500; they collect rainwater from the guttering and reuse it to flush the loos and to fill the washing machine. Adrian has monitored its performance as well: "When it's working, it supplies about 60% of our household water. I calculate it's saving us about £7 a week, which I am delighted with. The two tanks together hold 7,500 litres and it takes just an inch of rain to fill them up. There was one three-week spell last summer when they ran dry: then the heavens opened and they filled up in just 20 minutes!" There is an overflow system under the basement which collects ground water from the French drains as well as surplus from the rainwater tanks. It has two pumps, a main one and a mains/battery back-up, which raise the surplus water up to the house drainage system.

Above ground, the house holds fewer surprises but is finished to an immaculate standard throughout. It's largely a brick and block construction with beam and block ground floors and timber I-joists at first floor level. Adrian and Beverley remain as enthusiastic about their project as ever, despite having been going hard at it for over three years. With their energy bills lower than ever, the rewards will keep on coming.

## FACT FILE

**Names:** Adrian and Beverley Thurley
**Professions:** Pilot and social worker
**Area:** Cambridgeshire
**House type:** Six-bedroomed plus granny annexe
**House size:** 460m²
**Build route:** Self-managed subcontractors
**Construction:** Masonry
**Finance:** Abbey National
**Build time:** Oct '01-March '04
**Land cost:** £200,000
**Build cost:** £327,000 plus garage at £32,400
**Total cost:** £559,400
**House value:** £800,000
**Cost/m²:** £710

**30%**
**COST SAVING**

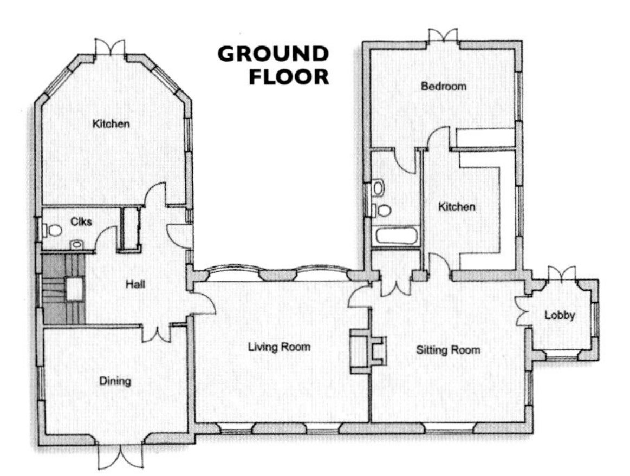


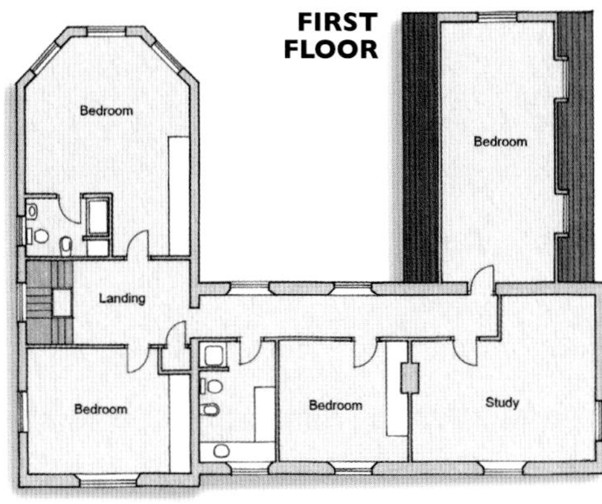


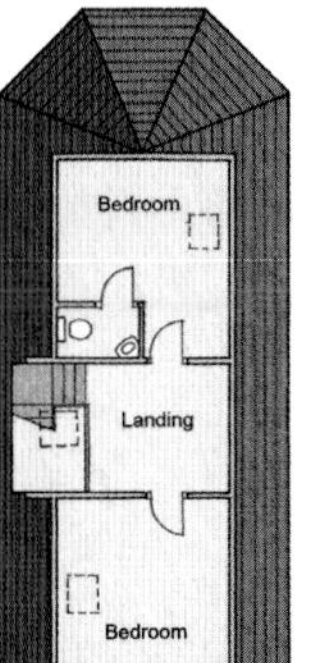


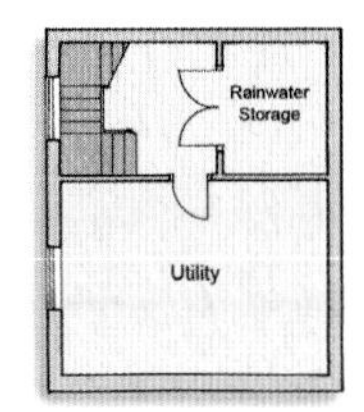


## FLOORPLAN

**The U-shaped design contains six bedrooms over four storeys, in addition to a granny annexe.**

## USEFUL CONTACTS

Adrian's website, **www.adethebuilder.co.uk** is a useful source of information on Adrian's project and self-building in general; Adrian and Beverley also have a lot of experience and knowledge to share; you can book B&B with them through the website; **Basement** ThermoneX Basements: 01204 559551; **Ground source heat pump** ICE Energy: 01865 882202; **Ventilation system** ADM Ventilation 01756 701051; **Builder** Steve Baldwin: 01638 507248; **Grey water recycling** Freewater UK: 0870 241 6964; **Kitchen** Pineland Kitchens: 01299 271143; **Fireplace** Fireplace Design Consultancy: 01889 500500; **Underfloor heating** Chelmer Heating: 01245 471111; **Plumbing** Baldwin & Wright: 01638 508457; **Carpenter** Terry Claydon: 07768 925831; **Designer** Roger Peyton: 01638 508360; **Grant** ClearSkies: 0870 243 0930 www.clear-skies.org

## ECO HOMEBUILDING FEATURES

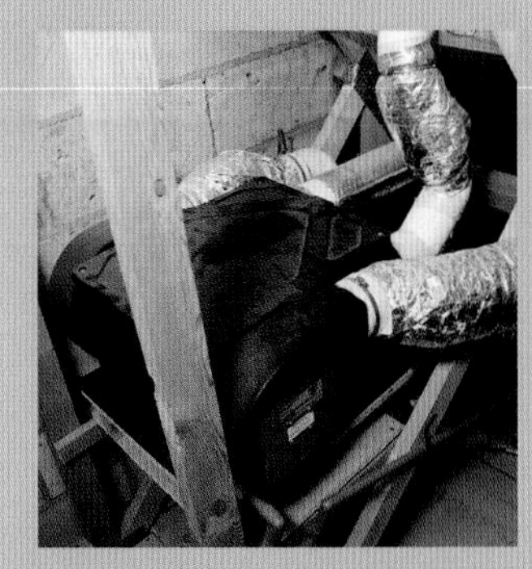

Adrian and Beverley's home is packed full of eco-friendly features that not only help to reduce the house's impact on the environment but keep down energy bills.

The ground source heat pump from Ice Energy heats the water up to 55°C, to be stored in a large pressurised cylinder. "It's been fascinating to see it working," says Adrian. "I was led to believe that it drew its heat from the ground, which remained at a constant 10°C all year around. But in fact the temperature in the garden loop has changed dramatically through the year, peaking at 21°C in summer and going as low as -1°C in winter. Yet even at -1°C, the heat pump still manages to extract 3°C or 4°C of heat from the pipes. For every one kilowatt of power used in running the system, we get between three and four kilowatts of heat out of it. That's the magic of heat pumps." It wasn't cheap, costing around £8,000 to install, but part of this cost was offset by a Clear Skies grant of £1,200.

The basement houses the rest of Adrian's eco features, including a whole house heat recovery and ventilation system (left) and a grey water harvesting/recycling system, which supplies around 60% of the household's water from the rain, used for the washing machine, flushing loos and so on. It cost around £2,500.

# VISION OF THE FUTURE

Richard and Sian Liwicki have built a contemporary, low-energy farmhouse in the walled garden of their Oxfordshire vineyard.

WORDS: DEBBIE JEFFERY
PHOTOGRAPHY: NIGEL RIGDEN

The existing walled garden and bothy shaped the design of the new house, which is built from a combination of stone-clad blockwork and timber framing, with glass shaded by a brise soleil overlooking the garden.

**The open-plan living area is cleverly demarcated with a change in ceiling heights. With so much wood on site, the Liwickis chose a woodburning stove with a back boiler for their sitting room, which is used during spring and autumn to provide hot water.**

When Richard and Sian Liwicki sit down to savour a well-earned glass of wine, they have the added satisfaction of knowing that the tipple they're drinking is their own, produced from vines visible from the windows of their newly constructed farmhouse.

Bothy Vineyard continues an ancient tradition of wine making in the Vale of the White Horse, Oxfordshire, where the warm sandy soil and mature vines help to produce some award-winning wines; but, surprisingly, the Liwickis are relative newcomers to running their own vineyard.

"Richard's a keen wine buff, and enjoyed volunteering at our local vineyard as a way to unwind in the fresh air after work," recalls Sian. "When the owners mentioned that they were retiring, Richard jumped at the chance to buy Bothy Vineyard. He negotiated a reduction in his hours at Oxford University in order to make the change, and I also gave up work to help run our new business.

"We're both scientists and have a great love of plants, as well as wine, so it seemed like the ideal choice – although it was also a challenge. Grape growing in England is a very labour-intensive activity, but we are lucky enough to have a faithful band of friends and volunteers who help and support us."

The couple have two daughters – Sasha, six, and Zoë, four – and recognised that the vineyard would be an idyllic place to raise a family, located as it is in a peaceful rural hamlet. Planning permission had previously been granted for a traditional-style house with an agricultural tie to be constructed on the green belt site, and the Liwickis approached Richard's old school friend, architect David Wylie, to talk over their ideas for building a new family home.

"WE WANTED TO BUILD A LOW-ENERGY HOUSE THAT WOULD BE VISUALLY EXCITING AND MAKE THE MOST OF THE VIEWS"

They wanted a more contemporary design which would remain sympathetic to its setting, and would offer the privacy they needed as a family while remaining connected to the vineyard itself. David responded by ensuring that the new house would effectively become part of an existing walled garden on the edge of the vineyard, and arranged the property as a series of private spaces facing into the garden. The planning officer was enthusiastic about the sensitive proposal and, as no objections were received, the design was approved under delegated powers.

"We'd been living in a barn conversion about seven miles from the vineyard, which we sold before renting during the build," Richard recalls. "We learnt a great deal from our previous homes, and were keen to avoid wasted corridor space. We also wanted to build a low-energy house that would be visually exciting and would make the most of the wonderful views. David came up with a design which links the house to the garden wall and an old stone bothy on the site, and it seemed like the perfect solution to all our needs."

**Stainless steel worktops have proved to be extremely practical and hard-wearing.**

'I OFTEN COOK EXTREMELY SPICY AND PUNGENT FOOD, SO WE DECIDED TO ISOLATE THE KITCHEN AND INTRODUCE A RESTAURANT-STYLE SWING DOOR'

The Edwardian bothy – previously used for wine making – was in poor structural condition and needed to be rebuilt. It was given a new first floor and now acts as the vineyard shop and office, which connects to the two-storey house via a single-storey flat-roofed link tucked down behind the garden wall.

This lower section of the building contains the kitchen and dining room, which overlook the garden through a wall of glass. In the two-storey section of the house, the dining and living areas combine to form an L-shaped room beside a double-height conservatory, which faces south to make the most of the light and warmth from the sun.

"Even before you start using new technologies like solar panels it's important to site a house correctly, insulate it fully and plan everything as sensibly as possible," says Sian. "Our house has been designed to trap heat from the low sun during spring, winter and autumn days, with shading above the glass to reduce the glare of the high summer sun. We also have a massive curved blockwork wall in the conservatory which is rendered in clay and acts like a radiator, absorbing heat and releasing it slowly."

The building is highly insulated and draught-proofed, and benefits from a natural ventilation system, which was especially designed by David Wylie to provide cooling during summer days. Solar panels on the slate roof warm the family's water, PV cells generate electricity from the sun, and a weather station has been installed to monitor and measure changes in the internal and external environment.

The north and east elevations of the house are timber framed, and have been clad in a distinctive rain screen of untreated cedar, with stone from the site facing the remaining cavity blockwork elements. "It was actually more expensive to recycle our own stone than to buy it in from a local quarry, because of the labour involved in cutting and preparing the different-sized blocks," explains Richard, "but we had a fantastic stonemason and the weathered finish was worth the extra effort."

With David Wylie based in London, an Oxford architectural practice was appointed to project manage the 14-month build, which was completed by a small and enthusiastic local building contractor, keen to learn about the environmentally friendly building principles and materials involved.

"To a certain extent we were all learning together, but our builders constantly talked to us and paid close attention to detail," says Sian. "The ➤

**The conservatory is naturally cooled and ventilated using the VAX (Ventury Air eXtraction) system.**

"IT WAS ACTUALLY MORE EXPENSIVE TO RECYCLE OUR OWN STONE THAN TO BUY IT IN FROM A LOCAL QUARRY"

**The master bedroom bridges the house. It features a round window overlooking the private garden and an oriel window which projects out to provide views of the vineyard.**

biggest problem was a delay caused by the German supplier of the glass curtain walling, which held everything up and cost us additional rent. We did try to buy locally whenever possible, and chose renewable materials such as bamboo and linoleum flooring, but sometimes it was impossible to find what we wanted in the UK."

Richard and Sian spent a great deal of time working with David Wylie on the design, discussing the way they like to live. They have reproduced certain features which they admire in other houses, such as the round windows, conservatory and double-height spaces, and the result is a practical and unusual family home.

Slit windows in the master bedroom overlook a void above the sitting room – enabling the couple to keep a watchful eye on the children from upstairs – and window seats offer inviting places from which to admire the views of the garden and the vineyard to the north-west.

The integral double-height conservatory is a dramatic multi-purpose space which is ideal for relaxing, drying clothes and growing plants. "It opens directly into the sitting room, which makes this area feel much larger," Sian explains. "Our converted barn had an open plan kitchen and living area, but I'm from south-east Asia and often cook extremely spicy and pungent food, so this time we decided to isolate the kitchen and introduce a restaurant-style swing door.

**Many of the best elements of contemporary design are in the details, such as the floor-to-ceiling glass screen in the shower and the shadow gaps, which replace skirting boards.**

"This is a working farmhouse, and at harvest time I prepare meals for large numbers of people, so it was important to have a hard-wearing kitchen and a practical utility with sinks and storage as well as a north-facing pantry," continues Sian. "This area has been built to the rear of the bothy shop, and connects directly into the kitchen, so I can walk through the house to serve in the shop and then return to domestic life. The entire house was devised so that we can run a business and retain our privacy at the same time. It's a concept which works extremely well and has made living here an absolute pleasure." ●

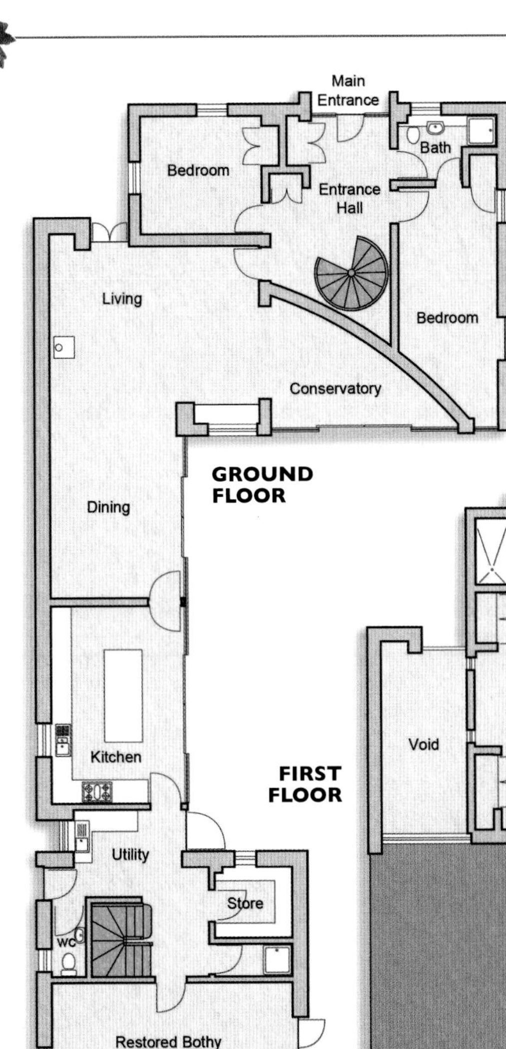

The restored bothy now contains the vineyard shop and connects to the two-storey part of the house via a single-storey glazed link.

# FACT FILE

**Names:** Sian and Richard Liwicki
**Professions:** Vineyard owners and assistant registrar at Oxford University
**Area:** Oxford
**House type:** Four-bedroom detached
**House size:** 300m$^2$
**Build route:** Building contractor
**Finance:** Private
**Construction:** Stone-clad blockwork, timber frame, cedar cladding, slate roof
**Warranty:** Architect's certificate
**Build time:** Jan '05-March '06

FIRST FLOOR

Bath
Void
Bath
Landing
Bedroom
Master Bedroom
Void
Void

# FLOORPLAN

**On the ground floor, the kitchen and dining room sit in the single-storey section of the house and overlook the garden through a glass wall. In the two-storey section, the living and dining rooms are open plan and lead into the double-height conservatory. Two bedrooms are situated to the rear. Upstairs are two more bedrooms, both with en suites. The office lies at the opposite end of the first floor and is accessed by a staircase from the ground floor utility.**

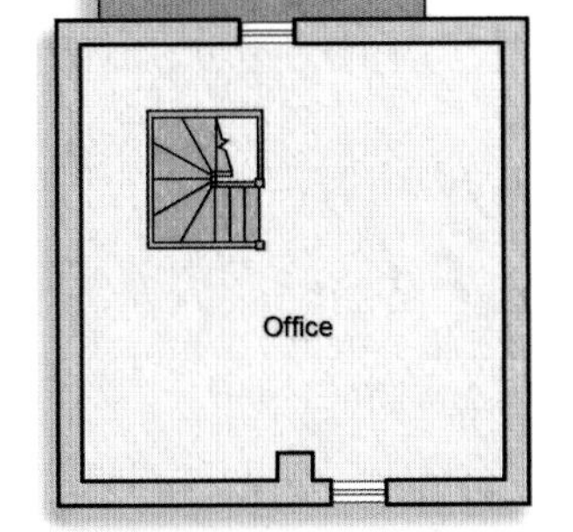


# ECO FEATURES

- **Solar gain/orientation**
- **Heat store**
- **Natural ventilation**
- **Solar panels (thermal)**
- **PV arrays**
- **Weather station**

# USEFUL CONTACTS

**Bothy Vineyard:** 01865 390067 www.bothyvineyard.co.uk **Concept architect** Wylie Associates: 020 8265 7620 www.davidwylieassociates.com **Project architect** Oxford Architects LLP: 01865 329100 **Builders** WG Carter Ltd: 01865 864626 **Staircase** Spiral Staircase Systems: 01273 858341 **Bamboo and linoleum flooring** Classic Flooring Ltd: 01865 773535 **Round windows** 3D Aluminium Plas: 01865 881403 **Oak doors** (bothy) OJ Joinery: 01559 371571 **Ventilation louvres** Bovema (UK) Ltd: 01244 400401 **Cedar cladding** D Smith Joinery: 01865 821095.

The home uses a partially dug basement, and makes the most of its south-facing elevation with solar panels and a conservatory.

# RETIREMENT ECO-PLAN

John and Joan Barnes have built a new home for their retirement with the intention of reducing energy bills and providing a comfortable lifestyle.

WORDS: DEBBIE JEFFERY PHOTOGRAPHY: NIGEL RIGDEN

"In 1965, I had my own building company, and designed and built a stone-clad house for my family," says John Barnes, 75. "We had a village plot overlooking fields, and acquired even more land when a new bypass was built close by and I asked if I could buy some of the wasteland they had used for spoil behind our property. I bought myself a JCB and became quite expert at terracing our 1.5 acres, which Joan landscaped."

Over time, the couple began to find it increasingly difficult to take care of so much land, however, and decided to try and build a more manageable house – something which they had discussed over a number of years. Their son David is an architect, and company director of architectural practice The Genesis Design Studio. John and Joan asked if he would design them a new home in the garden which would be suitable for their retirement.

The new house has been designed to exploit the slight slope of the ground and the southern aspect. A partial basement was built to maximise and overcome the sloping site, and the mass of the brick and block building helps to control daily and seasonal temperature swings.

Built over three levels, the house appears to be a small single-storey bungalow from the entrance, with only a few small square windows punched into the masonry to avoid directly overlooking the Barnes' previous home. The rear of the house faces south, and has extensive glazing to make the most of the views and passive solar gain from the sun. Patio doors open onto a balcony at ground floor level, with all habitable rooms arranged on the south and fully glazed.

A two-storey-high conservatory is also incorporated on the southern side of the house, and door openings into this space allow warmth from the sun to work its way up into each level of the main building, with automatic temperature- and rain-sensitive opening vents in the conservatory to prevent overheating. A Velux window above the back staircase also draws warm air up and out of the building to cool it when necessary.

Ground floor living rooms benefit from the wall of glass and a full-length balcony, part of which is external, with a central section contained inside the conservatory. John and Joan use the basement conservatory as a sun lounge, which is overlooked by an internal balcony where they can sit and eat breakfast.

"On the whole we managed to agree about most things, although David would have liked us to keep the ground floor living space totally open plan," says John, "but we decided to separate the sitting room from the kitchen and dining areas to allow us to use the rooms independently." ➤

"WE WERE GETTING FRUSTRATED BY THE LACK OF PROGRESS AND, AS I'M A QUALIFIED JOINER, I DECIDED TO HELP THEM OUT"

Small square windows light the timber and glass staircase, and are the only windows on the north side of the house.

Ceramic tiles and beech laminate flooring help to reduce dust mites which could trigger Joan's asthma. Right: Patio doors open onto a balcony at ground floor level overlooking the basement sun lounge in the conservatory below.

The three bedrooms and two bathrooms utilise the full shape of the roof, and have high sloping ceilings and exposed glulam beams, with a strip of Velux windows which reach down to the floor — affording views south across the Wiltshire Downs.

When the Barnes had built their previous house, they had lived on site in a caravan, and John had undertaken all of the building work — even making the doors and windows in his workshop. This time, however, he and Joan were able to stay in their own home while the build progressed.

"We intended to employ builders and subcontractors to build the house for us, but it didn't quite turn out as planned," says John Barnes. "Our son-in-law did most of the brickwork, but by the time we got up to roof level, the builders were scratching their heads over how to bolt glulam beams together. We were getting frustrated by the lack of progress and, as I'm a qualified joiner, I decided to help them out. From there on I became more and more involved, and ended up doing quite a lot of the internal carpentry myself as well."

The Barnes' house is highly insulated and has a solar panel to assist with heating the hot water, which is expected to reduce costs by around 50 per cent. Zoned underfloor heating is used throughout, which allows lower water temperatures in some of the heating circuits and suits the highly energy efficient condensing boiler. Rainwater is collected from the roof in an underground garden chamber to be filtered and recycled for toilets, the washing machine and garden watering, saving the couple a significant amount on their bills.

A continuous ventilation system with heat recovery prevents heat loss, but also substantially reduces draughts and moisture content. This improves air quality which, coupled with the use of tiled and laminate flooring, reduces the occurrence of house dust mites which could trigger Joan's asthma.

"We started out wanting a smaller, more manageable property for our retirement, but have actually ended up with a larger house than before," he says. "The layout allows us a certain amount of flexibility, and although we currently use the basement level as an office and workshop with a conservatory sunroom, this could easily become bedrooms or a self-contained unit, depending on our needs in the future. I'm no spring chicken – although I do still feel like one occasionally – but our self-build project was certainly an interesting challenge at my time of life." ●

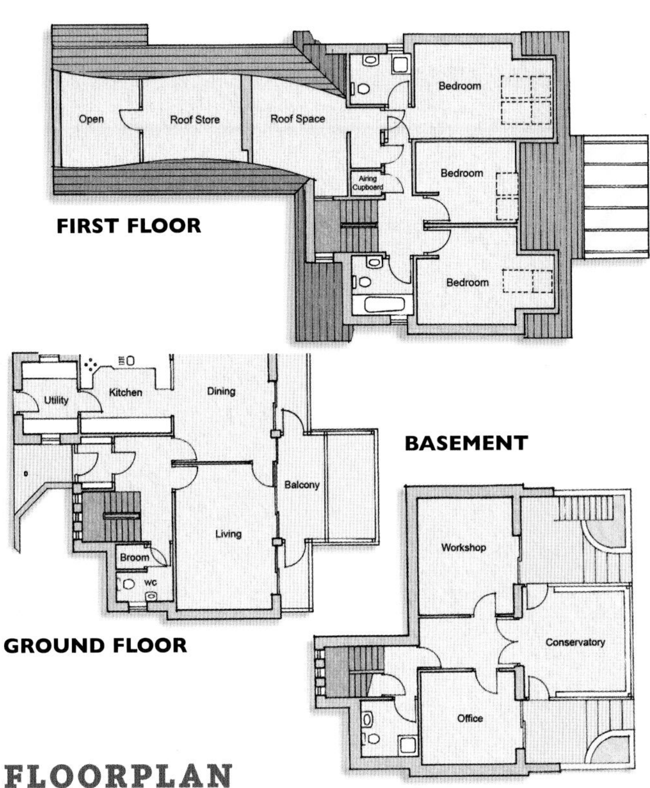


## FLOORPLAN

**The lower ground floor is currently a workshop and store, with a conservatory seating area, but has been fitted with a shower and could be used for bedrooms in the future. On the ground floor, the kitchen is open plan to the dining room and balcony beyond, with a separate sitting room and a WC and utility. Three bedrooms and two bathrooms have been accommodated in the roof space.**

## FACT FILE

**Names:** John and Joan Barnes
**Professions:** Retired
**Area:** Dorset
**House type:**
Three-storey, three-bedroom house
**House size:** 184m² + 14m² garage
**Build route:** Subcontractors and DIY
**Construction:** Brick and block walls, clay pantiles
**Warranty:** NHBC Solo for Self Build
**Finance:** Private
**Build time:** July '00-July '02
**Land cost:** Already owned, approx. cost £90,000
**Build cost:** £219,000
**Total cost:** £309,000
**House value:** £380,000
**Cost/m²:** £1,106

**19%**
**COST SAVING**

**Cost breakdown:**

|  |  |
|---|---|
| Building regs | £1,000 |
| Structural engineer | £750 |
| Site set-up and preliminaries | £3,050 |
| Services | £4,250 |
| Substructures | £18,000 |
| Masonry walls and lintels | £20,000 |
| Precast floors and screeds | £3,300 |
| Structural timber | £5,000 |
| Roof structure and covering | £20,600 |
| Ceilings | £2,000 |
| Windows, glazing, rooflights | £11,500 |
| Doors and ironmongery | £8,800 |
| Internal partitions | £3,500 |
| Internal joinery and stairs | £4,500 |
| Flooring | £6,000 |
| Kitchen fittings | £7,500 |
| Underfloor heating | £6,500 |
| Ventilation | £2,000 |
| Sanitaryware | £2,000 |
| Plumbing and heating | £9,500 |
| Solar panels | £1,800 |
| Water recycling | £3,100 |
| Electrical | £7,300 |
| Rainwater goods and soil pipes | £2,500 |
| Plastering and decoration | £14,200 |
| Conservatory | £8,800 |
| Balconies | £2,000 |
| Wall tiling | £500 |
| Insulation | £3,000 |
| External works, landscaping | £35,300 |
| NHBC | £1,000 |
| **TOTAL** | **£219,250** |

## USEFUL CONTACTS

**Architect** The Genesis Design Studio: 01794 519333; **Structural engineer** Andrew Waring Associates: 01794 524447; **Windows** Swedish H-Windows Ltd 029 2052 2246; **Rooflights** The Velux Company Ltd: 01592 772211; **External doors** Allan Brothers Ltd: 01289 334600; **Internal joinery** Granton Joinery; 01202 841963; **Ventilation system** Baxi Ltd: 01772 693700; **Underfloor heating** Pipe 2000: 01268 759567, Enviraflor Ltd: 01268 759567; **Rainwater harvesting** The Green Shop: 01452 770629; **Solar Panels** Sola Sense: 01792 371690; **Kitchen** Trade Kitchens (Dorset); 01747 850990; **Guttering** Dales Nordal System: 0115 930 1521; **Glulam beams** Structural Timbers Ltd: 01275 832724; **precast concrete cills** Forticrete; 0870 903 4015; **Bricks** Taylor Maxwell: 01962 835800; **General materials** Bradfords: 01935 813254; **Groundworks** Tercon Construction Ltd: 01202 676940; **Elctrical** BF Keane Electrical Contractors Ltd 01794 301481; **Mechanical** R Shepherd and Partner: 01794 512445; **Roofing** Shire Roofing Salisbury Ltd; 01722 329462.

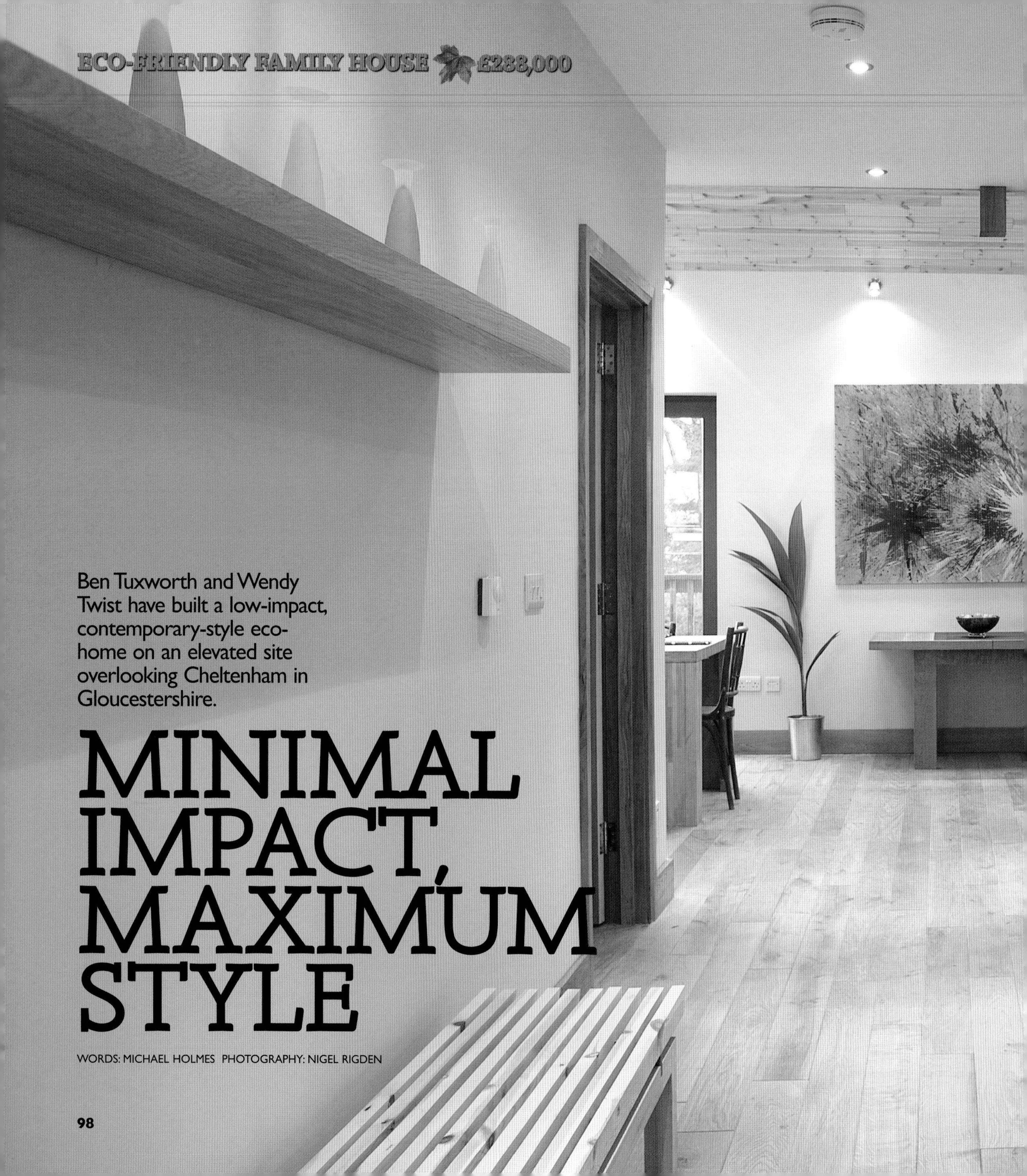

Ben Tuxworth and Wendy Twist have built a low-impact, contemporary-style eco-home on an elevated site overlooking Cheltenham in Gloucestershire.

# MINIMAL IMPACT, MAXIMUM STYLE

WORDS: MICHAEL HOLMES PHOTOGRAPHY: NIGEL RIGDEN

The elegant staircase made from laminated oak with glass balustrading and oak handrails joins all three levels.

**The three-storey house looks like a single-storey property from the front elevation, thanks to the wraparound deck and the way it is sunk into the site.**

Ben Tuxworth and Wendy Twist are a brave couple. In June 2003 they set out to build their dream family home on a largely DIY basis, equipped with a great deal of ambition but very little idea of how long it would take, how much it would cost and the enormous amount of energy the project would consume. "Had we really understood what we were taking on, we would probably never have started out," admits Ben, who works for a charity that advises businesses about environmental responsibility.

As if they needed any more complications to add to the stress, Ben and Wendy also agreed to let television cameras from Discovery Home & Leisure follow their build from beginning to end, telling the true story of their successes and their failures, in the series How to Build a House.

"When we started out we were convinced that we would be in by Christmas 2003 and that the house would cost us around £60,000," explains Wendy, who took on the role of project manager. "Not because we had worked out a budget, but because that was all the money we had left after buying the plot, and because that was when Ben, who took a six month sabbatical to work on site, had to go back to work."

When Christmas 2003 came, the house was still without a roof and much of their £60,000 budget had gone, together with their early optimism. "The groundworks were far more expensive than we had expected," recalls Ben.

"THE RESULT IS A SUPER INSULATED, AIRTIGHT SHELL THAT IS HIGHLY ENERGY EFFICIENT"

"Collapsing trenches caused by wet weather required masses of concrete, and an engineer-designed cantilevered slab – probably unnecessary – and a reinforced retaining wall, cost thousands."

Fortunately, Ben and Wendy's three quarters of an acre plot, which is on an elevated north facing wooded site, came with a three-bedroom timber shack. Although planning consent for the new house was only granted on the basis of replacing this dilapidated structure, the planners allowed it to remain in place until their new home was finished. Although far from luxurious, it at least provided Ben and Wendy and their two children, Isabel and Freddie, with somewhere reasonably comfortable and – more importantly – cheap to live, as it would be a further nine months before their new home was ready for them to move in.

So why did the build take so long? "We chose timber frame construction because it is meant to be fast and energy efficient," explains Ben. "However, much of the speed and efficiency of this form of construction comes from ➤

**The main living areas are on the first floor – entered via a raised walkway – and are filled with natural light from large areas of glazing and by banks of rooflights above the mezzanine level.**

**The open-plan kitchen has inexpensive IKEA units combined with worktops made from marine plywood.**

the fact that the frame is manufactured in a factory and assembled on site in a matter of days. Because of poor access and a steep slope, we felt it would be very difficult to get a factory-made frame onto site and so we decided to build one ourselves on site instead." This decision was fine in principle, and perfectly viable, only Ben and Wendy attempted it with no accurate drawings, no framing experience, and very little skilled help – for most of the build they had assistance from just one carpenter, plus the specialists they brought in for the groundworks, plumbing, electrics and heating.

One advantage of making the frame on site, however, was that Ben and Wendy were able to create a structure to their own exacting specification. This was a hybrid version of the closed panel 'breathing wall' design, 300mm thick, made using Masonite engineered timber I-beams, clad either side in strand board, and filled with cellulose insulation made from recycled newspaper. The roof is of a similar design, only with 500mm of insulation. The result is a super-insulated, airtight shell that is highly energy efficient. So low is the heat requirement that the property can be heated by a single ground source heat pump. This clever piece of engineering, which Ben and Wendy sourced from German manufacturer, Viessmann, heats the house by cooling the ground around it, via large heating coils buried about 1.5m down. Powered by electricity, it provides 4kW of usable heat for every 1kW of electricity it consumes. "It is a form of solar heating," explains Ben. "The ground remains at a constant temperature all year round just a few feet down. We extract this heat cost-effectively to power our underfloor heating, and to provide hot water, at very little cost. The payback period, compared to a conventional oil-fired central heating system, is around 12 years, but the system will last a great deal longer."

"COLLAPSING TRENCHES CAUSED BY WET WEATHER REQUIRED MASSES OF CONCRETE, AN ENGINEER-DESIGNED CANTILEVERED SLAB AND A REINFORCED RETAINING WALL COSTING THOUSANDS"

**The mezzanine floor provides a second living area, filled with light by two banks of Velux rooflights.**

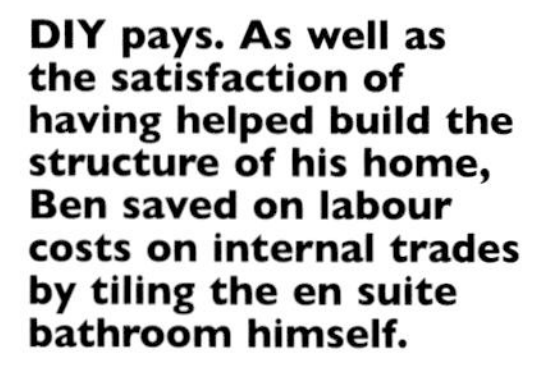

**DIY pays. As well as the satisfaction of having helped build the structure of his home, Ben saved on labour costs on internal trades by tiling the en suite bathroom himself.**

It is not just the structural frame of Ben and Wendy's house that is highly ecological. The exterior is clad in a combination of natural lime render, reinforced by horsehair, and horizontal oak boards, sawn from a tree felled from the site at the beginning of the project. The roof is covered in slate, which can be recycled, while inside, most of the finishes are timber, from sustainable forests.

Building an ecological house is not just about using 'green' materials that produce minimum pollution in their manufacture: it begins at the design stage, with basics such as orientation and fenestration, which should maximise use of daylight and passive solar gain, while minimising heat loss. "We worked with architect Charles Cox from a firm called Heath Avery," explains Wendy. "The house is partially built into the slope of the site which helps shield it from the elements, but also helps to reduce the visual impact of the front elevation, which is also broken up by the timber wraparound decking. To maximise the amount of daylight from the south, we included large banks of Velux rooflights and placed the living rooms upstairs, largely open plan, with a soaring vaulted ceiling and a mezzanine level. This arrangement also enabled us to make the most of the views over Cheltenham and the surrounding countryside. The bedrooms, which require less light, are on the ground floor."

Getting the project finished required several visits back to the bank manager who fortunately obliged, thanks to the fact that the project has proved a great success financially, as well as in design terms. Ben and Wendy may not have taken the fastest self-build route, but by project managing rather than using a main contractor they saved themselves a fortune, and ended up with exactly the house they wanted. "We have still got to complete the landscaping next spring, and we haven't got a penny left, but it is a fantastic house," says Wendy. "The project took a huge amount of energy, but it has all been worth it!" ●

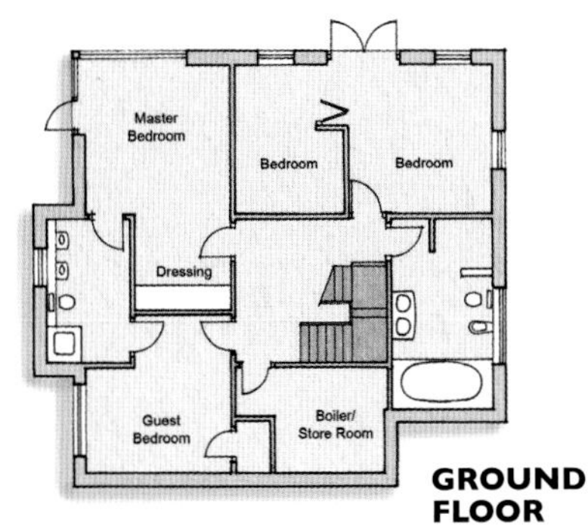


GROUND FLOOR

"THE HOUSE IS PARTIALLY BUILT INTO THE SLOPE OF THE SITE, WHICH HELPS SHIELD IT FROM THE ELEMENTS AS WELL AS HELPING TO REDUCE THE VISUAL IMPACT OF THE FRONT ELEVATION"

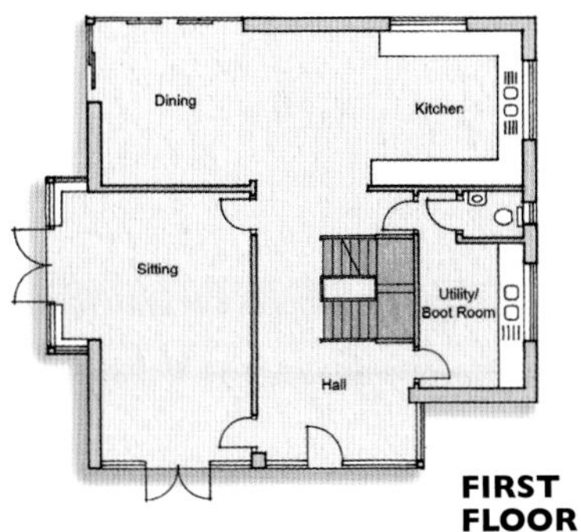


FIRST FLOOR

MEZZANINE

## FLOORPLAN

**The house is partially built into the sloping site. The main entrance is at first floor level, reached by an elevated deck walkway. The room layout makes the most of the views across Cheltenham by placing the living rooms on the first floor and the bedrooms below. A mezzanine level above the kitchen breakfast room provides a second living space, lit from above by a bank of Velux rooflights. Unusually, the master bedroom shares its en suite facilities with the guest bedroom.**

## FACT FILE

**Name:** Ben Tuxworth and Wendy Twist
**Profession:** Strategy director for ecological consultancy and homemaker
**Area:** Gloucestershire
**House type:** Brick boathouse
**House size:** 320m$^2$
**Build route:** Architect and contractor
**Construction:** Masonry and steel with piled foundations
**Warranty:** Architect's Certificate
**Sap rating**: 97
**Finance:** Private
**Build time:** June '03-Oct '04
**Land cost:** £248,000
**Build cost:** £288,100
**Total cost:** £536,100
**House value:** £725,000
cost/m$^2$: £900

**26%**
**COST SAVING**

**Cost breakdown:**

| | |
|---|---|
| Fees (incl. design) | £14,000 |
| Groundworks | £55,500 |
| Timber frame | £20,000 |
| Warmcell insulation | £6,200 |
| Roofing | £9,500 |
| Windows and ext. doors | £24,000 |
| Staircase and balustrades | £12,000 |
| Electrics and plumbing | £17,000 |
| Underfloor heating | £5,000 |
| Ground source heat pump | £6,400 |
| Dry-lining and decorating | £13,300 |
| Floor finishes | £9,500 |
| Second fix joinery (materials) | £5,700 |
| Kitchen and bathrooms | £10,500 |
| Decks and walkway | £23,000 |
| Landscaping | £5,000 |
| General labour | £33,000 |
| Avnex hub | £3,000 |
| Scaffolding | £5,500 |
| Beam & block floor and screed | £2,000 |
| External finishes | £7,700 |
| **TOTAL** | **£288,100** |

## USEFUL CONTACTS

**Architect** Heath Avery: 01242 529169; **Concrete** RMC: 01932 568833; **Pre-cast beam and block flooring** Rackham Housefloors: 01924 455876; **Timber frame** Panel Agency: 01474 872578; **Glulam/LVL** AJ Laminated Beams: 01284 828184; James Donaldson & Sons: 01592 752244; Finnforest (UK) 020 8420 0777; **Plywood** UPM Kymmene Wood: 01628 513300; **Decking** Timber Decking Association: 01977 712718; Howarth (Timber Importers) Ltd: 01469 532300; Distinctive Landscapes: 01623 647317; **Information on building with wood** Wood for Good: 020 8365 2700; **Recycled newspaper insulation** Warmcell: penycoed.construction@btinternet.com; **Windows and external doors** Rationel Windows (UK) Ltd: 01869 248181; **Roof windows** The Velux Company: 01536 510020; **Underfloor heating** Nu-Heat: 01404 549770; **Heat pump** Viessmann: 01952 675000; **Multi-room media system** AVNex: 01223 237700; **Dry-lining** Sasmox/Panel Agency: 01474 872578; **Timber flooring** Taylor Maxwell Timber: 0113 274 4655; **Pine skirtings, architraves, door frames** SCA Timber Supply: 01782 202122; **Oak skirtings, architraves etc.** Timbmet: 01865 862223; **Construction of stairs and balustrades including glass** BC Joinery: 01242 254343; **Mosaics** Waxman Ceramics: www.waxmanceramics.co.uk; **Central vac system** Beam Central Vacuum: 01386 849000; **Passive stack ventilation system** Passivent: 0161 962 7113; **Plasterers for lime render** G&J Haddow Plasterers: 07989 600525; **External lime render** Traditional Lime Company: 01242 525444; **Acoustic floor** Sound Service: 01993 704981; **Sanitaryware** www.bathstore.com; **Internal doors** STP Joinery: 01663 744030; **Roof tiles –** Aaron Roofing Supplies: 01452 731133; **Plumbers** www.heating2order.com: 0870 756 7820; **Engineer** O'Brien & Price: 01242 237227.

# HIDDEN GEM

Mark and Liz Ward overcame their fair share of problems in building a beautiful new traditional-style timber frame home in a spectacular woodland setting.

WORDS: DEBBIE JEFFERY PHOTOGRAPHY: JEREMY PHILLIPS

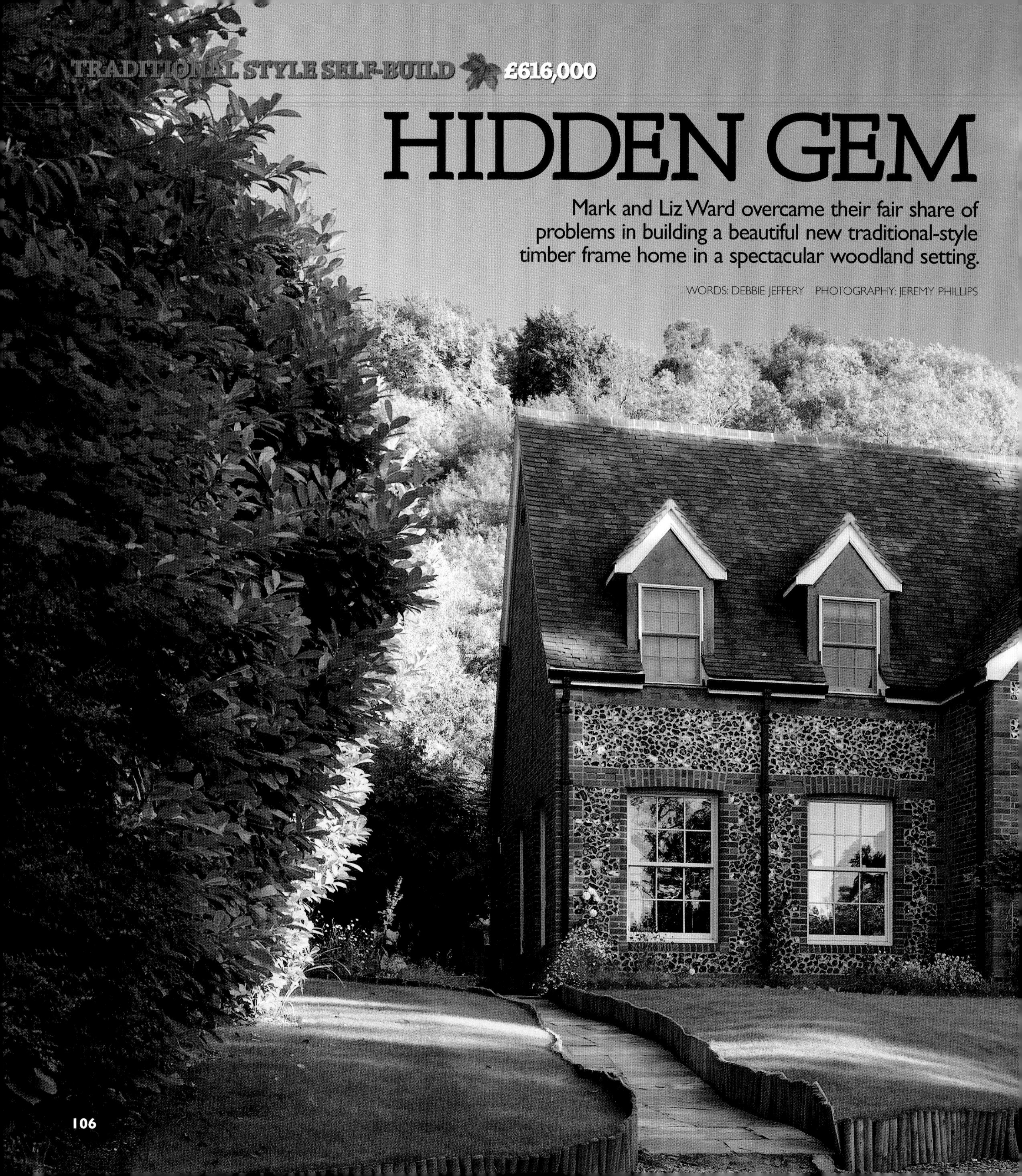

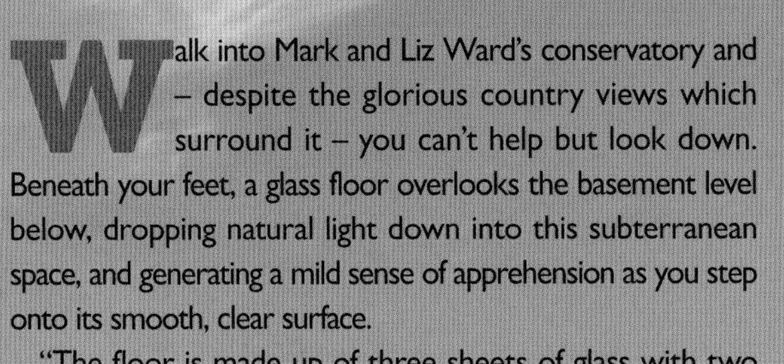

Walk into Mark and Liz Ward's conservatory and – despite the glorious country views which surround it – you can't help but look down. Beneath your feet, a glass floor overlooks the basement level below, dropping natural light down into this subterranean space, and generating a mild sense of apprehension as you step onto its smooth, clear surface.

"The floor is made up of three sheets of glass with two layers of laminate in between, so it's incredibly strong. It was installed by a specialist company, but visitors still often try to walk around the edge," smiles Mark, who enjoyed introducing quirky features into the family's new home. "The same thing happens when they step onto the glass balcony from our bedroom!"

Glass plays a vital role in the Wards' house design, adding a touch of glamour and ensuring that valuable natural light finds its way down into their basement. This entire lower ground floor level has been designed just for fun, with a studio, a games room and a large heated swimming pool which the family can enjoy using all year round.

"When Liz's father died, we inherited his 1950s bungalow in Wendover, standing on a magnificent three-quarter-acre site

backing onto woodland," says Mark. "There were so many happy memories connected to the place that we just couldn't sell, so instead we decided to replace the old bungalow with a new house which would make the most of the wonderful views. The last thing we wanted was for these views to be marred by a swimming pool though, so a basement seemed like the perfect way to gain the extra space we wanted."

The Wards were aware that the bungalow stood in an Area of Outstanding Natural Beauty, adjoining a Site of Special Scientific Interest, and that a replacement house would need to respect the sensitive nature of this setting. Ultimately, they settled on a traditional cottage-style property, with walls of red brick and knapped flint, and a rustic entrance porch complete with reclaimed oak posts.

"We knew that we wanted a timber framed house, and met with Potton at one of the self-build shows," says Liz. "They were able to offer a design service which merged our ideas with elements from their own range of houses to make something really unique. It took two years to get planning approval, partly because incorporating a basement took us over the planners' recommended size guidelines — even though it would be completely hidden from view."

Planning permission for a two storey house was initially obtained and a separate application was then made which included the basement level. Once this had been approved, the Wards and their two children – Hannah and Robbie – were able to sell their previous 1970s home one mile from the site, and moved into a purpose-made log cabin, which they built at the bottom of the garden.

"WHEN THE BUILDER WENT BUST – TAKING £30,000 OF OUR MATERIALS MONEY WITH HIM – WE HAD NO CHOICE BUT TO STEP IN TO MANAGE THE PROJECT"

**The sitting room is double sided and also serves the entrance hallway and where a woodburning stove is fitted in the brick inglenook.**

"We didn't want to waste money renting a house, but a caravan or mobile home wouldn't have been large enough to take our furniture and a baby grand piano," says Liz. "In the end, we had a three bedroom cabin built for us which had a vaulted ceiling and a huge living room with a verandah under the overhanging pitched roof. We even used some windows taken from the demolished bungalow, and lived there for almost 16 months. Once we'd moved into the new house we sold the log cabin for £15,000, so effectively it didn't cost us a penny."

The first stage of the build meant engaging an experienced company for the basement design and build package. The site stands in an area of chalk, and excavations involved digging out four metres for the basement, with a further two metres for the deep end of the centrally positioned swimming pool. More than ten tons of structural steelwork supports the house above the basement.

Potton took just three weeks to erect the timber frame, and the Wards then employed a local building contractor to complete the house, including building the external brick and flint skin and tiling the roof with the handmade clay tiles, which help to give the property its distinctive and traditional appeal.

"We had no intention of getting involved with the build, and were happy to sit back and watch," says Mark, "but when the builder went bust three months after the timber frame was finished – taking £30,000 of our ➤

A structural glass floor in the conservatory drops light into the basement, where windows have been kept to a minimum to meet planning conditions

materials money with him – we had no choice but to step in to manage the project and do as much of the work as possible. The idea had been to build using the profits from our previous house sale, and to end up mortgage free, but in the end we had to take out a mortgage in order to complete the project."

**The kitchen/breakfast room is from Magnet (01296 489095) and contains steps which lead down to the basement level**

The Wards employed a number of subcontractors directly and tackled a great deal of the work themselves, including decorating and laying floors. Everyone lent a hand, and the children spent their half-term holiday fitting insulation. This involvement brought the family closer together and has meant that they all take huge pride in the finished house.

"We had to watch the budget much more closely, but we were determined not to scrimp and still chose the best quality materials we could afford," Mark says. "The Andersen windows are a particular feature. Internally, they are finished in maple, which matches our woodwork and the main staircase."

This staircase leads up to a large landing, which doubles as a reading area, and makes far better use of the available space than creating another guest bedroom. Downstairs, back-to-back fireplaces serve both the hallway and living room, whilst stairs lead down from the spacious kitchen/breakfast room to the basement level below.

Here, the pool is flanked by glass panels, and Liz had the idea of building a fish tank into the wall between the pool and the neighbouring games room. "It's quite surreal swimming in the pool and looking out at the fish," she says. "We've also positioned a table-tennis table beneath the glass balcony in our bedroom, so you can look down and watch the children playing below. People wonder whether there's a privacy problem with so much glass, but this is a family home, so it really isn't an issue."

Like their basement, the Wards' hi-tech gadgetry is also carefully hidden from view. A ground-source heat pump, with coils of pipe buried in the garden, warms the swimming pool water as well as supplying the underfloor heating and other domestic hot water around the house, and there is a ventilation and heat-recovery system which ensures that the basement remains warm and dry all year round.

**The £20,000 swimming pool is flanked by glazing, with a picture frame-style fish tank inset into one internal wall**

"We think it's great that you can build a house which looks like a traditional brick and flint country cottage but has the advantages of modern technology," says Mark. "From the outside you simply wouldn't know that there's a basement here at all, and we can enjoy all the pleasures of an energy-efficient, low-maintenance home, with luxuries like a heated pool, without imposing on the beautiful site." ●

## FACT FILE

**Names:** Mark and Liz Ward
**Professions:** Chartered surveyor and sports coach
**Area:** Buckinghamshire
**House type:** Four bedroom detached house with basement level
**House size:** 418m$^2$
**Build route:** Builder, subcontractors and DIY
**Finance:** Private and mortgage
**Construction:** Timber frame clad in brick and flint
**Warranty:** Zurich
**Build time:** Aug '04 – Nov '05
**Land cost:** £500,000
**Build cost:** £616,000
**Total cost:** £1,116,000
**Current value:** £1,500,000
**Cost/m$^2$:** £1,474

**26%**
**COST SAVING**

**FIRST FLOOR**

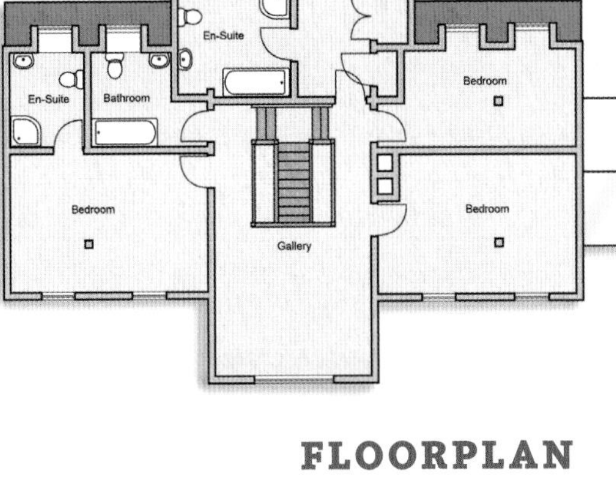


**GROUND FLOOR**

**BASEMENT**

## FLOORPLAN

**The basement level contains a games room, swimming pool and studio, with space for a sauna beside the separate shower and WC. The vaulted ground floor entrance hallway opens into the kitchen, sitting room, study and dining room, and a maple staircase leads up to four spacious bedrooms, a bathroom and two en suites**

## USEFUL CONTACTS

**Basement** ThermoneX: 01204 559551 **Timber frame** Potton: 01767 676400 **Windows and doors** Andersen Windows: 01283 511122; Sashtec: 01491 614498 **Structural glass and glazing** Glazeguard South West: 01823 337755 **Underfloor heating** Comfort Corporation: 01295 672280 **Ground-source heat pump** Ice Energy Heat Pumps: 01865 882202 **Swimming pool** Chiswell Pools: 01923 269822 **Plumbers** William Peters: 01296 331911 **Roofer** GM Roofing: 01296 331968 **Carpenter** Eddie Atkins: 01296 583127 **Brick and flintwork** Len Edmonds: 07909 870504 **Structural steelwork** Triangle Metal Works: 020 8886 0172 **General building supplies** E East & Son: 01494 433936 **Conservatory** The Conservatory Roof Co: 01525 853403 **Kitchen worktops** Granite Solutions Direct: 01753 528777 **Balcony balustrading** Q-Railing: 0800 781 4245 **Spiral staircase** Stairs Direct UK: 0870 814 7760 **Door furniture and ironmongery** The Brassware Shop: 020 8360 7771 **Maple staircase** C&A Associates: 020 8574 1193 **Kitchen sinks** Franke UK: 0161 436 6280 **Gutters and downpipes** Harrison Thompson & Co: 0113 279 5854 **Handmade bricks** Dunton Brothers: 01494 772111 **Internal glazing** Lordship Double Glazing: 020 8885 3702 **Pebble tiling** Island Stone: 0800 083 9351 **Reclaimed oak** Site 77: 01296 631717 **Flint and lime mortar** Old House Store: 0118 969 7711 **Air-handling system** Starkey Systems: 01905 611041 **Maple floorboards and Belgium bluestone** Fired Earth: 01295 812088 **Honed antique limestone floor and boulders** Artisans of Devizes: 01380 720007 **Jacuzzi baths** Windsor Bathrooms: 01782 717517 **Basins** Porcelanosa: 0117 959 7150 **Picture frame aquarium** Waterworld Aquatics: 01732 760991 **Decking** Briants of Risborough: 01844 343663 **Handmade roof tiles** Midlands Slate & Tile: 01902 790473 **Kitchen** Magnet: 01296 489095 **Woodburning stove** Morley Stoves: 01920 468001 **Building guarantee** Zurich: 01252 377474 **Kitchen appliances** Neff: 01908 328300 **Plasma TV and sound systems** Audio Vision: 020 8447 8288 **Garden swimming pond** Anglo Swimming Ponds: 020 8363 8548

To the south, a splash of acid-green render combined with horizontal timber cladding hints that this is no ordinary house. The existing exterior brickwork has been clad in StoTherm Mineral System insulated render for a fireproof and weather-resistant finish.

MP Alan Simpson and his wife Pascale Quiviger have turned a derelict lace mill in Nottingham into a sleek, stylish beacon of an eco home.

# A NEW SHADE OF GREEN

WORDS: DEBBIE JEFFERY PHOTOGRAPHY: NIGEL RIGDEN

MOVING FORWARD

**Cardboard tubes which run up behind the ground floor staircase like giant stems of bamboo are manufactured by Essex Tube Windings from 100% recycled fibreboard.**

It's not just Blue Peter presenters who are able to rustle up useful items out of old jam jars and sticky-backed plastic. MP Alan Simpson's unusual new home has been built using recycled cardboard tubes, broken bottles, straw and various other unlikely materials, but the end result is far from shoddy.

The once derelict and unwanted building, formerly a small-scale lace mill right in the heart of Nottingham's vibrant and historic Lace Market, is completely enclosed by other buildings and surrounded by all manner of hotels, bars and car parks – making access a logistical nightmare. This previously insignificant structure has been transformed into a contemporary and sustainable home for Alan, his French Canadian wife – the award-winning author and painter Pascale Quiviger – and their baby daughter Élie, who was born in January 2006.

Alan Simpson has been the Labour MP for Nottingham South since 1992. Born in Bootle, Liverpool in 1948, he has lived and worked in Nottingham for the past 30 years and is a leading campaigner on a wide range of issues about the environment and the economy. Voted the sexiest MP in Nottinghamshire, and dubbed by The New Statesman as 'the man most likely to come up with the ideas', Alan was determined to practise what he was preaching by proving that it needn't cost an arm and a leg to breathe new life into substandard and redundant buildings.

"In the past I had looked for older houses and done them up, but nothing quite on this scale," explains Alan, who met Pascale, 36, part way through the building project when he was temporarily living with his sister and her husband. "My plan was to find the most run-down, unpromising property that I could and turn it into an attractive, sustainable home for a realistic budget. Then a friend found this place for me, which was a real eyesore and

"MY PLAN WAS TO FIND THE MOST RUN-DOWN, UNPROMISING PROPERTY I COULD AND TURN IT INTO AN ATTRACTIVE, SUSTAINABLE HOME FOR A REALISTIC BUDGET"

**The open-plan kitchen is fitted with Homebase units and recycled woodblock worktops from Ikea, with Bosch Class A appliances. Marmoleum flooring – a natural product made of limestone, wood flour, resin, pigments, jute and linseed oil – is an ideal choice for people with allergies.**

ticked all the right boxes."

Alan paid £100,000 for the boarded-up derelict structure, which is hemmed in by other buildings and has no outdoor space. "People recoiled in horror when they walked inside and saw the knee-deep piles of pigeon droppings," recalls Alan, who invited long-term friend and local architect Julian Marsh of marsh:grochowski to come up with some cost-effective eco solutions to the numerous problems. "He's the most visionary and gifted architect I've ever come across, and has never knowingly designed a room full of right angles in his life."

The staircase had collapsed so that Alan couldn't even access the upper floor of the building before he bought it, and bare brick walls were all that remained of the original rhomboid-shaped structure. This has been re-

"GETTING LIGHT INTO THE BUILDING WITHOUT LOSING OUR PRIVACY WAS A BIG PROBLEM, AND ONE OF THE SOLUTIONS WAS TO DESIGN A FEATURE WINDOW FOR THE LANDING FROM RECYCLED WINE BOTTLES"

roofed and heavily insulated internally with Homatherm insulation slabs made from recycled wood chippings. Internal walls were constructed using a composite board formed from compressed straw, and even the sterling board flooring is more usually found boarding up broken shop windows. Roof-mounted photovoltaic cells provide electricity while a state-of-the-art ➤

Light drops down through the 1m x 4m polycarbonate rooflight to the rear of the house and passes through the sandblasted glass which forms a landing on the top floor.

Bathing becomes a shared pastime, with the steel Bette bath prominently positioned in the master bedroom

"I'VE NEVER HAD A PROBLEM WALKING OVER THE GLASS... SOME PEOPLE CAN BE CAUTIOUS ABOUT THE DROP BELOW"

WhisperGen Micro Combined Heat and Power system (the first in the UK) acts like a boiler but, ingeniously, it also drives a generator to produce a further 20% of the average household's energy needs. Surplus energy is exported back to the National Grid – resulting in cheques from the electricity company in lieu of bills.

"Being green doesn't have to be about wearing sandals and eating lentils – it can also be exciting. We wanted to make a feature out of the water-recycling stacks, and decided to label them so that people would know what they are," says Alan of the two bumps protruding from either side of the landing, which have been painted with the words 'brown' and 'grey'.

"Getting light into the building without losing our privacy was a big problem, and one of the solutions was to design an obscure feature window ➤

"BEING GREEN DOESN'T HAVE TO BE ABOUT WEARING SANDALS AND EATING LENTILS – IT CAN ALSO BE EXCITING"

for the landing from recycled wine bottles," he continues. "The builders said that, as a labour of love, they would make their own contribution by gathering together as many empties as they could after work on a Friday night!"

It was not possible to form any new openings in the exterior walls to the living space due to the close proximity of the neighbours, and so light steals have been built into the terrace above, with the glass top shafts lined in mirrored Perspex to allow daylight to be drawn down into the previously dingy kitchen area below. To the rear of the house the same problem applied, but this time a polycarbonate rooflight was inserted, below which a sandblasted glass landing enables natural light to drop three storeys down through the stairwell.

Such clever conceits ensure that every corner of the house is brightly lit. On the ground floor the layout is predominantly open plan, with a large living/dining/kitchen in which the structural steel supports of the original lace mill create an industrial flavour. Rearing up from this double-height space the minimal white staircase is backed by a wall of startling yet inexpensive recycled cardboard tubing — giving the impression of giant stems of fat bamboo growing uniformly up through the house.

**The rear of the house faces north onto an adjacent car park and has only two existing windows at the upper level, creating an unobtrusive façade which belies the building's exciting interior. A photovoltaic (PV) system on the roof converts solar radiation into electricity – producing clean, emission-free electricity for the home.**

The entire first floor is dedicated to the master bedroom suite and features a capacious steel bath, tiled in mosaics, jutting out into the sleeping area. This opens through glass doors onto a suntrap roof terrace which has been paved in tiles made predominantly from recycled glass.

On the top floor, two further bedrooms and a shower room are currently used for guests, but one of these will eventually accommodate Élie when she grows old enough to leave the cot in her parents' bedroom. By which time, the house will have started to pay for itself in terms of energy savings and will, in Alan's words, "tread lightly on the future." ●

## GREEN FEATURES

- Combined heat and power system (CHP)
- PV cells
- Rainwater harvesting
- Recycled wood chippings insulation
- Marmoleum flooring
- Recycled building materials

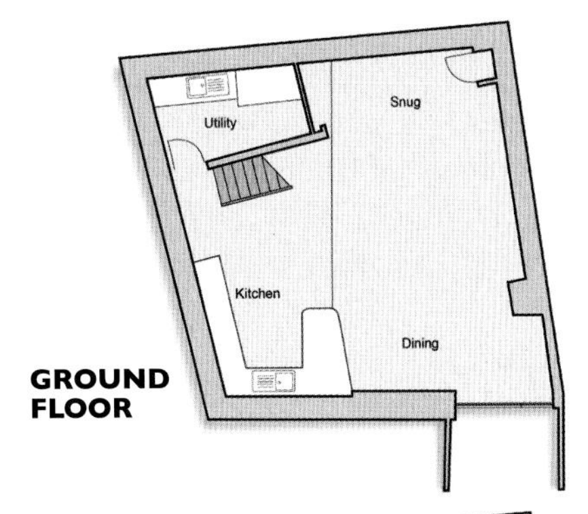


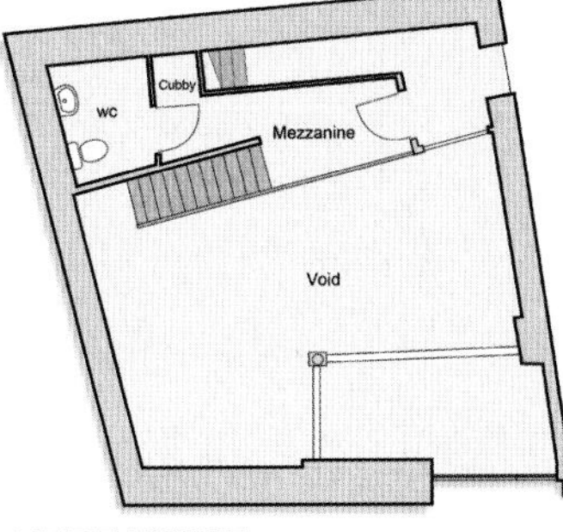


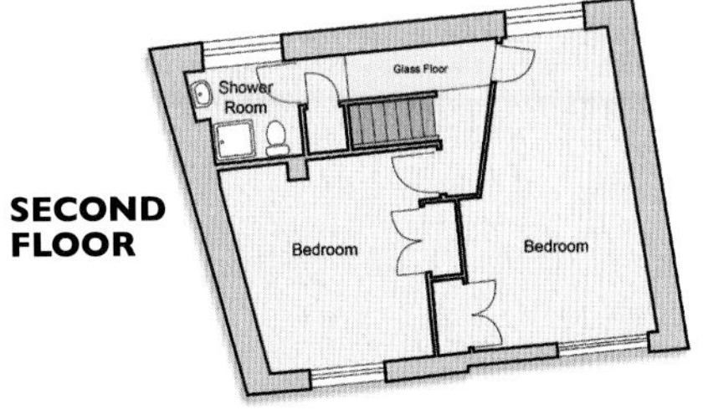


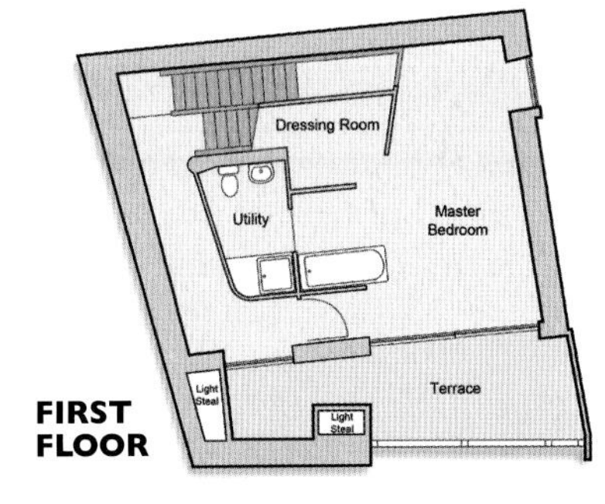


## FACT FILE

**Names:** Alan Simpson and Pascale Quiviger
**Professions:** MP and author/painter
**Area:** Nottingham
**House type:** Converted lace mill
**Build route:** Main contractor
**Finance:** Private
**Construction:** Brick/ masonry/steel
**Build time:** Aug 2004-Aug 2006
**Mill cost:** £100,000
**Build cost:** £200,000
**Total cost:** £300,000
**Current value:** unknown

## FLOORPLAN

**On the ground floor, the layout is open plan, with a large living/ dining/kitchen area, plus a small, separate utility room. The first floor is occupied by the master bedroom, while the second floor features two guest bedrooms and a shower room.**

**The derelict lace mill before work started.**

## USEFUL CONTACTS

**Class A appliances** Bosch: 01905 752640 www.bbt-uk.co.uk **Steel bath** Bette: 01789 262262 www.bette.co.uk **IDO dual-flush WCs** Construction Resources: 020 7450 2211 www.constructionresources.com **Polycarbonate rooflight** Danpalon: 01858 468323 www.danpal.com **Cardboard tubes** Essex Tube Windings Ltd: 01375 851613 www.essextubes.com **Builders** Framework UK: 0115 974 3734 **Waterwiser recycling system** Gramm Ltd: 01273 844899 www.grammenvironmental.com **Raindance showers** Hansgrohe: 0870 7701972 www.hansgrohe.co.uk **Glass floor and bottle window** Makers: 0115 941 9290 **TJI joists** Panel Agency Ltd: 01474 872578 www.panelagency.com **WhisperGen Micro Combined Heat and Power system** Powergen: 0800 068 6515 www.powergen.co.uk **Insulation** Rockwool: 01656 862621 www.rockwool.co.uk **Photovoltaic solar panel** Solar Century: 020 7803 0100 www.solarcentury.co.uk **Recycled glass tiles** Spartan Tiles: 01206 230553 www.spartantiles.com **Exterior insulated render** Stotherm Mineral System: 01505 333555 www.sto.co.uk

# NATURAL LIGHT

Peter and Mhairi Taylor have replaced their 1950s bungalow with a contemporary self-build incorporating cutting-edge construction techniques and eco features.

WORDS: BEVERLEY BROWN
PHOTOGRAPHY: DOUGLAS GIBB

It took just two hours to demolish the home Peter and Mhairi Taylor and their two sons had lived in for 17 years. But as the couple had plans to replace the 1950s chalet bungalow with something far more exciting, any feelings of nostalgia were replaced by the thrill of anticipation.

"The old house was poorly insulated and faced the wrong way anyway, so rather than spend money on it we decided to knock it down and build new – and we were keen to do it in an environmentally friendly way," explains Peter. ➤

Located in the former grounds of a 19th century mansion house on a green belt site near Edinburgh, the Taylors' property included a small but separate lodge house, destined to be their home for the duration of the build. Living on site was particularly valuable as Peter and Mhairi were determined to manage the project themselves.

**In some areas of the construction, steel framing was required to support the vast areas of low-emissivity double glazing. Photovoltaic solar cells have been integrated into the atrium to generate electricity.**

It's perhaps just as well the Taylors were living on site, as the man responsible for the design – Peter's brother-in-law and architect Lindsay Johnston – was on the other side of the world. An Irishman who studied in Scotland and now lives in Australia, Professor Johnston is internationally recognised in the field of low-energy, environmentally sustainable building design.

"We gave Lindsay our remit, which included lots of light, open spaces, and an indoor/outdoor lifestyle, and he sent back a model, along with a tape he'd recorded to explain his thinking behind the design," says Peter. He adds: "When we said we were delighted, Lindsay came over to talk in more detail and liaise with a local Edinburgh architect to submit planning etc."

Although the site is technically green belt, the project was a one-for-one replacement and on that basis, acquiring planning permission was straightforward – the only proviso being a height restriction to prevent the building overshadowing a nearby listed stone archway. It also helped that services, including sewerage, were already on site.

"FULL-HEIGHT WINDOWS WITH GLASS-TO-GLASS CORNERS MAXIMISE VIEWS TO THE FIRTH OF FORTH."

The design is both inspirational and deceptive: an elongated detached four-bedroom bungalow with a predominantly open plan layout over three floors. From the outside, the design pays lip service to a traditional one-and-a-half-storey style. But there's a contemporary twist in stainless steel gutters, dormer windows with glazed sides and flat roofs, decks that blur the distinction between the interior and the garden, and full-height windows with glass-to-glass corners, designed to maximise unobstructed views over open countryside to the Firth of Forth. Equally striking, the line of the steeply pitched roof extends to an adjoining conservatory at one end, where a double-height glazed ceiling incorporates photovoltaic cells to convert the sun's rays into electricity.

SIPs (structural insulated panels) construction forms the basis of Lindsay's design: a system chosen for its design flexibility, low wastage and superior insulating properties. Essentially a timber 'sandwich' with an insulating, dense and non-permeable centre, a SIPs house can reduce heating bills by as much as 60 per cent. It's also stronger than a conventional timber frame, eliminates condensation, and, once completed, is indistinguishable from any other house.

By the time Lindsay Johnston arrived in Scotland, Peter and Mhairi were busy researching geothermal heat pumps and obtaining quotes for everything from solar panels and whole-house ventilation systems to a garden water harvester. "At the same time we were also trying to find a builder, but having approached nine companies, none were interested, mainly because the non-standard design featured long glass walls supported by a steel frame," says Peter.

When BPAC in Fife accepted the project, the couple broached the possibility of using geothermal heating. "We'd read about it in a self-build magazine and were keen to incorporate it until we heard horror stories about equipment that didn't work," he says. "Then we went skiing in Norway and met an architect who showed us a Swedish heat ➤

**The central glass and steel staircase is the focal point of the house, leading from the basement up into the glass atrium.**

An impressive multi-fuel stove keeps the living room warm and cosy.

**The kitchen/ dining room has a gas-fired Aga and slate flooring.**

"AT TIMES, THE BUILDERS WERE RELUCTANT TO TRY NEW THINGS, BUT WE ALWAYS GOT THERE IN THE END"

pump available from a UK distributor and that convinced us to go ahead."

Based on the fact that the temperature underground remains at a constant 10-13°C, geothermal heating systems are either based on pipes sunk several hundred feet into the ground linked to a heat exchanger to provide domestic hot water and feed central heating radiators (in the Taylors' case, underfloor heating), or use a ground-source heat pump and pipes laid horizontally closer to the surface to harness the sun's energy. Not having the large surface area required for the second option, the Taylors chose to sink geothermal pipes into a specially constructed borehole. "We were eligible for a grant to offset the cost of installing the solar system and ground-source heat pump, but the application process is complicated and needs to be simplified," Peter comments.

Living on site, Peter and Mhairi were hands-on from the start, clearing the ground in preparation for the concrete foundations. One contractor did the under-build, basement, and beam and block floor, thereafter BPAC, the main contractor, took over. "There were so many decisions to make, it was never ending," says Mhairi. "In between sourcing products on the internet, going to specialist shows and exhibitions, and generally just talking to people, we cut and laid all the base insulation – 90mm thick and around 400m$^2$ of it – before the poured screed and underfloor heating could be laid. And we always made a point of tidying up the site at the end of each day, which gave us the chance to check progress."

All these things took a massive amount of energy. Mhairi and Peter also landscaped the garden themselves. "Like the house, it too evolved, depending on where the earth landed," she laughs. As the couple are keen gardeners, this challenge was one they relished, particularly as they managed to incorporate stone from their original home into the setting.

Gradually the house took shape. SIP exterior walls (1,670mm-wide panels) were strapped and sheeted with a 100mm cavity and 100mm blockwork. The roof is also SIPs, a system that doesn't require roof trusses or any other support that will reduce the amount of space. Spanish slate tiles give the roof a traditional finish. Steel framing was required in places to support vast, low-emissivity double-glazed windows and sliding patio doors. "We did think about using self-cleaning glass for the atrium above the staircase, but the company needed three months to test the ceiling material and we couldn't wait that long," Peter explains. The German-designed Schüco conservatory was built by Northern Tectonics, who also installed the photovoltaic solar cells in three arrays, facing east, west and south. The electricity is sold directly to the grid, with every unit generated credited by Good Energy.

For the next three years the project became a full-time occupation. "You do everything yourself because you know what you want," Peter reflects. "At times, the builders were reluctant to try new things, but we always got ➤

**A ground-source heat pump feeds the underfloor heating throughout the house. The tiles and trims are from Porcelanosa (0131 335 3883).**

"THIS WAS A VERY NON-STANDARD HOUSE, SO IT TOOK A LONG TIME TO BUILD AND INVOLVED A LOT OF RESEARCH"

there in the end. We were lucky in that we had relatively few problems overall. Some trades didn't turn up because they were too busy elsewhere, and the render company held us up for several months – but we're quite patient people really," he adds.

The Taylors' home is notable for its clean lines and a deceptively spacious interior that includes a basement, open plan ground floor with a dining hall, living room and kitchen/family room linked by huge sliding panelled doors, made from American white oak and designed by Mhairi. Upstairs are four bedrooms with coombed ceilings, and three bathrooms. Additional rooms have been worked into the design – a cloakroom off the hall, his and hers studies off the kitchen, and four stores, a washroom, laundry, plant room and hobby room, off the adjoining conservatory.

Equally impressive are the things you can't see, such as the loft-based Nuaire home ventilation system, which continuously inputs filtered fresh air from the outside. And the reams of cabling masterminded by their son Iain, a partner in Edinburgh-based Ezone Interactive (a website management system). Using 'flood wired' cabling, which caters for every conceivable eventuality, the main AMX controller has the ability to control virtually everything, although at present it's only configured for home entertainment. Control panels in each room mean that different music can be selected as an alternative to whole-house listening. The house also has Wi-Fi internet access and wireless lighting, while three zones (basement, conservatory and living room) have been wired for home cinema. "Based on the amount of cable he bought and what he was left with, Iain recently calculated we have 7km of cabling in total," says Peter.

Most of the Taylors' other choices are more visible, including a gas-fired Aga in the kitchen, Phoenix multi-fuel stove in the living room, Philippe Starck bathroom fittings, and Caithness slate floors in the kitchen and adjoining conservatory. The couple appreciate good design but prefer simple, clean lines, and their home reflects a preference for quality over quantity.

There are some unusual features too, such as the angular staircase in the dining hall, with a clear glass balustrade and stainless steel support. "We'd initially wanted something less fussy," says Peter, "but the legalities meant it was not to be." Upstairs, the master bedroom has French doors leading onto a north-facing balcony that provides superb views of the Forth estuary, while two adjoining bedrooms at the other end of the house each have double doors out onto a shared balcony that overlooks the conservatory.

"This was a very non-standard house, so it took a long time to build and involved a lot of research. Now that we're actually living in it, we love it," says Peter. "It achieved what it set out to – it's an outside/inside house. There are cosy corners, but opened up, it can be a big hospitable space." Would they do it again? "Absolutely, but perhaps a smaller country version." ●

## FACT FILE

**Names:** Peter and Mhairi Taylor
**Professions:** Founders of the Town House Hotel Company
**Area:** Edinburgh
**House type:** Four-bedroom contemporary self-build
**House size:** 500m$^2$
**Build route:** Selves as project managers with two main contractors
**Finance:** Private
**Construction:** SIPs with Spanish slate roof and rendered exterior
**Build time:** Oct '2002-June '2005
**Land cost:** Already owned

**Build cost:** £550,000
**Total cost:** £550,000
**Current value:** £1.25m
**Cost/m$^2$:** £1,100

**56%**
**COST SAVING**

## FEATURES

**● Whole-house ventilation system ● Ground-source heat pump ● Structural insulated panels ● Large expanses of glazing ● PV cells**

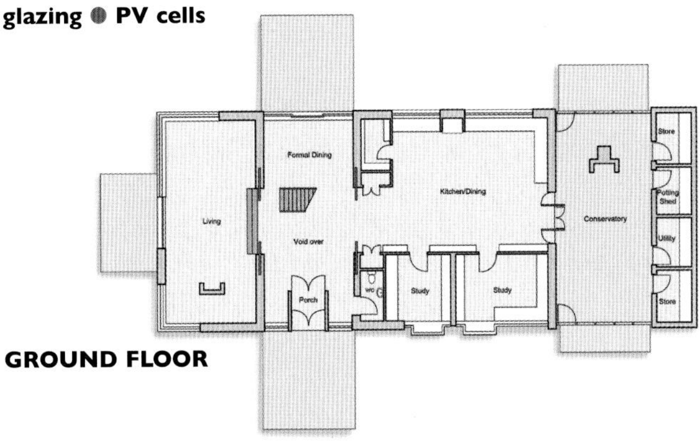


**GROUND FLOOR**

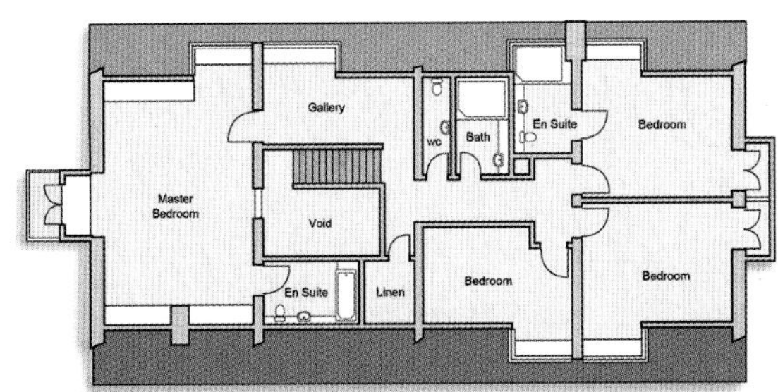


**FIRST FLOOR**

## FLOORPLAN

**On the ground floor, the kitchen/dining room leads through huge sliding doors to a more formal dining area from which a staircase leads both up to the first floor and down to the basement. Upstairs are four bedrooms and three bathrooms.**

### REBUILD OR REPAIR: WHAT'S GREENEST?

Weighing up the decision about whether to renovate a run-down house or demolish and rebuild involves considering a range of differing factors, such as value, design, planning policies and so on. If the primary motivation behind the project is to create a greener house, then the homeowner has to balance out the competing elements: the embodied energy in the manufacture of the new house vs the comparatively low energy consumption of a new dwelling. Research suggests that homes built even as recently as the 1960s consume twice as much energy each year as ones built in the 1990s – and regulations have become much more stringent since then. As it takes between 5-10 years for a home to use as much energy (in heating etc.) as it does to build it, it's fair to conclude that there is a good argument to say that the greenest solution for our housing stock is to knock some of it down and start again. Architect of the Taylor house, Lindsay Johnston, paid strong consideration to these factors during the project. "Adaptive reuse of existing building stock is a primary strategy towards sustainable cities. To 'throw away' the existing buildings is to throw away much of the energy consumed in their making."

### USEFUL CONTACTS

**Architect** Professor Lindsay Johnston: www.rivertime.org **SIPs** BPAC (now SIPs Industries): 01383 823995 **Geothermal heat pump** Ice Energy: 01865 882202 **Conservatory** Northern Tectonics: 01307 461700 **Underfloor heating** Rehau: 01989 762600 **Ventilation system** Nuaire: 029 2085 8441 **Cabling for smart home system** Ezone Interactive: 0131 225 6622 **Rainwater harvesting** Freerain: 01636 894906 **Joinery** Holyrood Joinery: 0131 652 1082 **Flooring** Caithness Slate: 01847 831980.

# ELEGANTLY EFFICIENT

The traditional style façade of Kieran and Lisa Robinson's new self-built home hides contemporary style interiors and the latest energy-saving features.

WORDS: HEATHER DIXON
PHOTOGRAPHY: JEREMY PHILLIPS

After living in the wings of a converted 17th century former girls' school for nearly two years, putting up with draughty windows, high ceilings and old radiators, Kieran and Lisa Robinson were ready for a few modern conveniences.

There were some advantages to their existing home, however. The school, which had been painstakingly renovated and developed by Kieran and his parents over five years, overlooks the Esk Valley near Whitby in North Yorkshire, and the Robinsons knew they would be hard pressed to find another home in a similar location.

**The kitchen from Rational features a Neff induction hob, Elica cooker hood and Corian work surface with inset sink.**

"WE COULD HAVE TAKEN ENERGY CONSERVATION FURTHER, BUT THERE REACHES A POINT WHEN YOU NO LONGER GAIN"

"We wanted the light and space we had enjoyed in the apartment, but we also wanted a low maintenance, energy efficient home – and beautiful views," says Kieran. "There was only one solution."

The Robinsons decided to build their own home in the grounds of Carr Hall at Ruswarp, positioning it just a few hundred metres from their last home, but higher up the hillside at the entrance to a small development of luxury houses, built by the Robinson family with retired couples in mind.

Architect drawn plans had already been approved by Scarborough Borough Council's Whitby office for a traditional two-storey brick house on a site positioned against the gently sloping hillside, but Kieran and Lisa wanted to include as many energy efficient concepts as possible into the build. They redrew the plans based on a timber frame to include solar panels to help provide hot water and photovoltaic cells to help reduce their electricity bills, plus a mechanical ventilation system with heat recovery.

"The aim was to create a house which was different to other people's," says Kieran. "Lisa and I had always wanted to build our own home but, as a property developer, I also wanted to find out just how energy efficient I could make it. By installing the most up-to-date concepts and experiencing the results first-hand, we could establish just how successful the whole energy saving business could be."

Although Lisa and Kieran had to reapply for planning permission to include solar panels in the roof, their scheme was passed without objection by Scarborough Borough Council.

"David Green, the planning officer, was quite keen on the idea and very supportive of what we were trying to achieve, which made the process that much more straightforward," says Kieran.

As well as incorporating energy saving systems, the Robinsons also wanted to create a home which was family friendly, as Lisa was expecting their first baby at the time of the build. They rearranged the internal layout of the rooms to create a 185m$^2$ home which was far less 'traditional' and better suited to their lifestyle. They then presented their drawings to timber frame manufacturers EBS, near Halifax, who immediately took the project on board.

"They took a rough sketch drawing from us then did their own full scale drawings with structural calculations," explains Kieran. "They were able to supply an NHBC certificate at the end of the project. The beauty of a timber framed house is that you alter them as you want. There are two structural walls downstairs, but no walls support anything upstairs so there is freedom to arrange rooms without any restrictions."

Kieran chose a standard Mansard truss roof with a distinctive deep overhang, which was then felted and tiled with Brazilian slate.

Four electric solar photovoltaic panels, each providing 0.25kw of electrical power, were installed in the roof, including that of the detached garage. Three solar water panels heat a 350 litre tank of water for everyday use, losing less than three degrees of heat in a day if the water is left standing.

"We usually have a full tank of hot water a 60°C plus, although after three

"FRESH AIR COMING IN IS NOT ONLY WARMED BY THE AIR GOING OUT, BUT ALSO FILTERED TO REMOVE DUST AND POLLEN"

days of bad weather I turned the hot water on from the boiler as the temperature was down to 35°C," recalls Kieran. "We have had it as high as 79°C in the tank."

The three panel system, a Kit 800, is made up of Vitosol 100 flat plate collectors installed on an inset roof system and require installation by a Viessmann key installer in order for the warranty to be valid. The Robinsons also installed an A-rated condensing boiler with weather compensating controls, which is housed in an upstairs cupboard.

"We had to have double glazed sash timber windows because we live in a conservation area, and we also had to match the style of the original school building which is Grade II listed," says Kieran. "It wasn't an issue for us. We would have chosen wooden frames anyway because they are more in keeping with the rest of the house."

The windows didn't require air vents because of the heat recovery/ventilation system from Viessmann. "The system effectively takes all the damp stale air out of the bathrooms, toilets and kitchen and puts it through a heat exchange box in the roof," says Kieran. "Fresh air coming in is not only warmed by the air going out, but also filtered to remove dust and pollen."

The system also measures the outside temperature and compensates for extreme cold or heat within the heat exchange system, to ensure that the temperature inside remains at a constant level. "We've now been in for almost a year and the system seems to be 98 per cent efficient," says Kieran.

The Robinsons wanted to use stone on the exterior of their house to complement the natural stone of the main school building, choosing a locally quarried Eskdale stone from nearby Egton, with each piece dressed and given a uniform Herringbone pattern on the external face. It took one man 12 weeks to complete the task single-handedly but the result, says Kieran, was well worth the wait.

The result is a home which, from the outside, is very much in keeping with its location and surrounding buildings but which, internally, is just about as energy efficient as you can get.

"We could have taken the energy conservation even further, but there reaches a point when you no longer gain," says Kieran. "As it is, we are probably saving at least 30 per cent on our fuel bills as a result of everything we have installed. I worked it out that in the course of 66 days, the cost of our central heating and hot water by gas amounted to just 16 pence a day — that's £1.18p a week. Brilliant! I'm quite sure that many of these things will become standard issues as the need for a reduction in fuel consumption becomes critical. Now we've lived in the house for a while we know it is a success. We just wonder how we managed without it before."

**"I WORKED IT OUT THAT IN THE COURSE OF 66 DAYS THE COST OF OUR CENTRAL HEATING AND HOT WATER BY GAS AMOUNTED TO JUST 16 PENCE A DAY"**

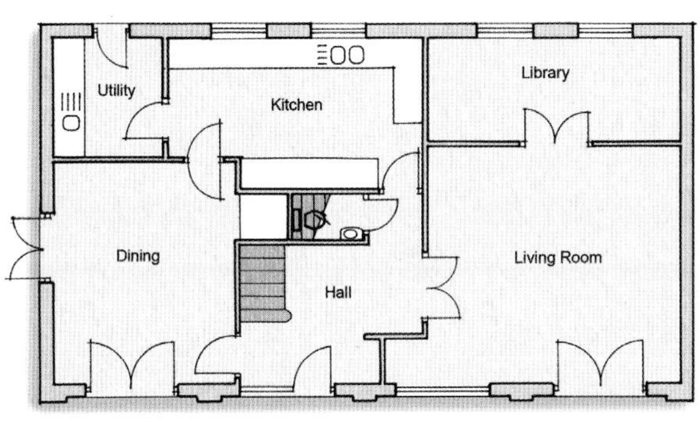


**GROUND FLOOR**

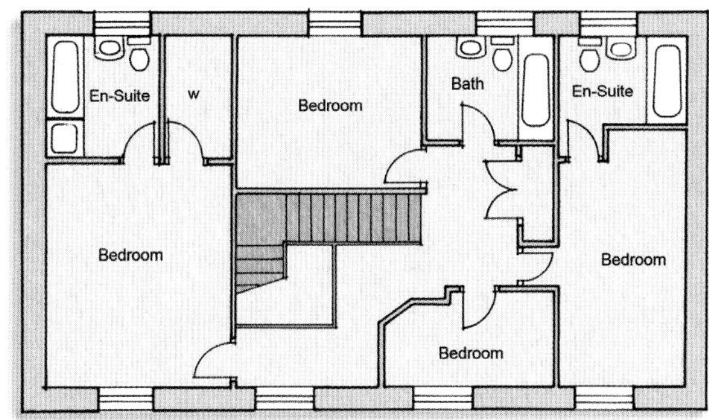


**FIRST FLOOR**

**Names:** Lisa and Kieran Robinson
**Professions:** Employee with Yorkshire Coast Homes and property developer and partner with Vito Tec Solutions
**Area:** North Yorkshire
**House type:** Four-bedroom detached
**House size:** 186m²
**Build route:** Self-managed with subcontractors
**Construction:** Timber frame with stone exterior and slate roof
**Warranty:** NHBC
**Finance:** Yorkshire Bank
**Build time:** 10 months
**Land cost:** £80,000
**Build cost:** £200,000
**Total cost:** £280,000
**House value:** £390,000
**Cost/m²:** £1,075

## 28%
**COST SAVING**

**Cost breakdown**

| | |
|---|---|
| Timber frame | £25,000 |
| Front doors and all windows | £8,000 |
| Solar panels for hot water | £740 |
| Solar Photovoltaic (electric) panels Vitovolt 300, with all connections, controllers etc | £7,285 |
| Viessmann Vitodens 100 24Kw weather compensating condensing boiler | £3,525 |
| Hot water tank: Viessmann Vitocell-B 100 (350 litre dual coil cylinder for solar applications) | £1,764 |
| plus safety kit | £700 |
| Underfloor heating | £400 |
| Heat recovery system (Viessmann Vitovent) | £2,585 |
| Roof: Grey slates: | £12,000 |
| Herringbone-faced coursed stone: | £85/yd² |
| Sanitaryware – | |
| En suite: Jacuzzi J tower walk-in 1400 | £2,290 |
| Sinks: Pizzi-Ginori of Milan 500 range | £151 each |
| Taps: Visio basin monoblocs by Pegler | £120 each |
| Bathroom – Bath: Jacuzzi Argus bath with six jets | £1,995 |
| Sink: Avanti in white Corian | £470 |
| Mirror: Schneider | £432 |
| Flooring: Amtico MTP232 tread plate lead | £70/m2 |
| Rational Kitchen, including Neff induction hob, oven, microwave, dishwasher, fridge/ freezer, Elica cooker hood and Corian work surface with inset sink and Franke tap: | £16,000 |
| **TOTAL** | **£200,000** |

## USEFUL CONTACTS

**Architect** Malcolm Tempest Ltd: 01677 450777; **Windows** Beacock & Sons Joinery: 01724 854370; **Solar panels, boiler, hot water tank** all products are manufactured by Viessmann Ltd: 01952 675000; **Plumbing installation, hot water, central heating and underfloor system solar panels and PV panels all installed by** Vito Tec Solutions Ltd: 01947 810242; **Roof** Dodds: 01377 272777; **Timber frame** EBS Northern (Environmental Building Solutions): 01422 327335; **Stone** Eskdale Stone Ltd: 01947 820821; **Sanitaryware, Kitchen, Amtico flooring** purchased from: HS Interiors: 01642 243403; **Amtico flooring was fitted by** Anthony Hodgson Flooring specialist: 01642 724305; **Insulation** – Seconds and Co Insulation products: 01544 260501.

# PART OF THE SCENERY

WORDS: CAROLINE EDNIE PHOTOGRAPHY: ANDREW LEE

**Neil Hammond has made the most of a stunning location by building an energy-efficient house that complements its setting beautifully.**

If the essence of successful homebuilding does, in fact, boil down to location, then Neil Hammond has hit the jackpot. After a journey through the preternaturally beautiful landscapes of Glencoe, Glen Nevis and Glengarry, the road ends at the historic Pictish site of Glenelg, across the bay from the eastern tip of the Isle of Skye. And it's here that Neil Hammond decided, unsurprisingly, that he'd like to build his dream home.

But we're not talking any old dream home blithely imposed upon the Highland landscape. No, Neil Hammond has helped create a home which is a triumph of eco-friendliness, that doesn't so much respect the dramatic landscape setting as worship it. In fact, such is the harmonious effect of the house in its hillside

**The use of locally-sourced materials – in addition to the turf roof – help the home blend in to its magnificent Highland setting.**

**Neil has chosen to leave all the wood untreated. "We used linseed oil on the frame, and what we've done in areas such as under the deck and in the cladding is to use larch because of its high resin content."**

location overlooking Glenbeag, that it only really becomes apparent that there is a house here at all halfway up the long and winding steep track that leads to Neil's door. And when it does become apparent, it is obvious that this timber and glass house with turf roof is not your average new Highland home.

"When I bought into a piece of land up here, which is basically a big environmental restoration programme involving reforesting, I wanted to build a house using my own skills in renewable energies and architect Neil Sutherland's design," explains Neil Hammond. "Rather than go with the traditional white kit bungalow type which are peppered over the landscape, usually built in the cold and frosty glens, ours sits perfectly into the hillside. This is why it's on five different levels, and also why it's got a grass roof. And the situation of the house, and the fact that it is a long building with a large south facing glazed terrace, also maximises solar gain, as well as allowing views in all directions."

The building itself is a hybrid of kit house and timber frame, post and beam, and timber panels. The timber frame, in a ridge beam arrangement, appears to the fully glazed south. The timber panel element is located in the north wall, which has little glazing, and acts as a shelter against the prevailing winds. Externally the house is clad in larch, with high efficiency, low emissivity double glazing, grass roofing and steel flashings.

In terms of materials, Neil has, as far as possible, sourced them locally. "We've used timber that's all sourced in Scotland. The Douglas fir main frame came from Glen Affrick, which is just over the hill, in three lorry loads. It was all machined here, then we built the frame at a yard down the bottom of the hill and brought the pieces up and bolted them together. The

"THE WHOLE IDEA BEHIND THE HOUSE WAS THAT IT WOULD BE SELF-SUFFICIENT IN TERMS OF ENERGY"

decking is Scottish oak, the screens and the doors were all made in Dornoch, by a company called Treecraft, and even the galvanised steel connectors were manufactured in the Highlands."

The recycled newspaper insulation is pivotal to the effectiveness of the house in terms of making the most of the renewable energy sources which power the house. As Neil explains: "The whole idea behind the house was that it would be self-sufficient in terms of energy. To achieve this you have to reduce your energy demand, and in order to do this you must insulate it way beyond normal levels." So there's 300mm thick Warmcell insulation in the roof, and 250mm thick insulation in the walls.

The energy self sufficiency is achieved by a 6kW wind turbine, and 1kW solar photo-voltaic panels. Neil, who is involved in the renewable energy industry, had previously installed a mast which collected data on wind speeds and solar gain in order to match the turbine with the amount of energy the house would require. As a result he has devised a system whereby the turbine and solar panels charge up a main battery which power the lights and computers and so on – although Neil has fitted out the house with low energy appliances to minimise the demand. All the surplus power is then linked into a water storage tank, and when this reaches a certain temperature, it circulates into the underfloor heating system.

Neil admits, however, that there were some compromises he did have to make along the way. For example, polystyrene insulation was used under the ➤

**Neil left his kitchen design up to Adrian of Adrian's Wood Workshop in Drumnadrochit, and while the kitchen appears to fully fitted, all elements are freestanding.**

The master bedroom has become, in effect, Neil's office – the elevated level enhances the glorious south-facing views .

"THE HOUSE TAKES A QUANTUM LEAP FROM THE DAYS WHEN SUPER ENERGY EFFICIENT HOUSES HAD TO BE UGLY"

floor, although it is green in that it doesn't contain any cfcs. Then there are the concrete foundations, which Neil admits aren't the most environmentally friendly thing to use, but "part of the philosophy is to store the solar energy through the day, so you need some kind of thermal mass in the building. Concrete was the only choice open to us really," explains Neil.

A cold and dark January precipitated the decision to install a high efficiency gas boiler as backup, "which cuts in when everything else has expired." And since Neil had a few teething problems with the turbine last winter he will have to wait to find out if the renewable energy sources can prevail in their own right as he believes they will.

Despite these compromises, however, Neil's vision of "new standards for living but also a nice living space," have certainly been realised. The Hammond house takes a quantum leap from the days when super energy efficient houses had to be ugly at the same time, and has succeeded in creating a simple yet stylish contemporary statement. Exterior forms merge seamlessly into the hillside setting, and interior spaces are defined by a light filled airiness and a clean, rational yet homely feel.

Internally, natural stone and timber flooring pave the way. A variety of local hardwoods make up the doors, architraves and skirtings as well as the stairs, in-built furniture and the south-facing kitchen area, which is one of the best spaces of the five-bedroom, three-level house. "The kitchen is big because in every house I've ever had, I've spent most of my time in there," says Neil.

Of course, no self-respecting eco home is complete without a composting toilet. However, if you're thinking crude hole in the ground,

**The turf roof houses the solar panels, while, in aesthetic terms, it allows the building to fit in with its environment.**

then think again. This is the latest state of the art model from Canada and according to Neil is completely hassle-free. In addition, Neil also reluctantly installed a conventional WC as he wouldn't have been granted a building warrant without one.

In fact, according to Neil, the whole planning issue proved difficult. "Getting planning was extremely difficult. The original planner's report was damning because the house didn't fit in with what they had in mind, namely the kit bungalow in the glen. When we first went to committee we had to lobby really hard – it was just lucky that we had a very good local councillor at the time, he was very supportive. Historic Scotland also objected because it is near an important Pictish archaeological site and they didn't want the house to be seen from the road. But what we've found is that the people visiting the site are really interested in our house as it is so unusual."

Rather than use a contractor, Neil opted to manage the construction work himself. "I built the house myself with friends and different people who became involved. Kenny the joiner was the main man; Adrian built the wood interiors; Barry the plumber was brilliant, and we also had a German Zimmerman craftsman too." Although the team certainly had their work cut out for them since there were around 16 external corners to construct, the experience seems to have been an enjoyable team effort.

Everybody got involved when it came to cutting the turfs and laying the roof. Neil is justifiably delighted with the grass roof and affords it the last word. "I think the roofing is really pleasing aesthetically. It adds weight to the building, so it's never going to go anywhere. And the real beauty is that it's silent when it rains." Which in Scotland's famously intemperate climate, is no small consideration.

## FACT FILE

**Name:** Neil Hammond
**Profession:** Property Developer
**Area:** NW Scotland
**House type:** Five-bed detached
**House size:** 140m²
**Build route:** Architect & contractor
**Construction:** Post and beam with turf roof and timber cladding
**Sap rating:** 100
**Finance:** Private + mortgage
**Build time:** June '00-Aug '01
**Land cost:** £15,000
**Build cost:** £120,000
**Total cost:** £135,000
**House value:** £150,000
**Cost/m²:** £900

**10%**
**COST SAVING**

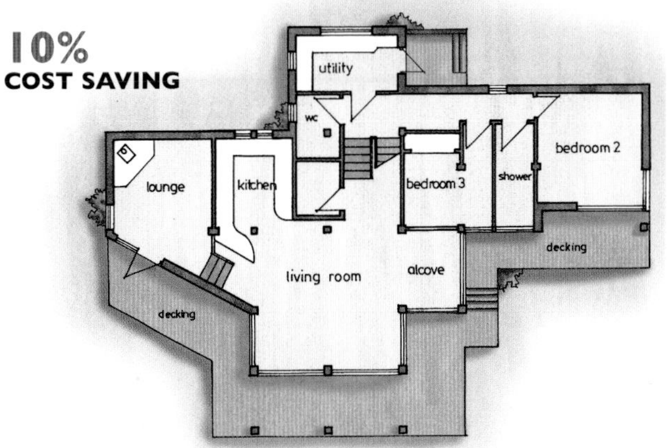


## FLOORPLAN

**The house is built into the contours of the hillside, hence it has five different levels.**

## USEFUL CONTACTS

**Architects** Neil Sutherland: 01463 709993; **Main Contractor** Natural House Construction: 01599 522288; **Construction Frame** built by Natural House Construction, Glenelg 01599 522288; **Roofing** Turf on Platon membrane, Proctor Group: 01250 872261; **Windows and doors** Treecraft woodwork: 01862 810021; **Kitchen manufacturer** Adrian's Woodworkshop, Cannich 01456 476268; **Bath and Shower** Bathroom Planet, Inverness; **Compost Toilet** Sun-Mar Corp, Canada, supplied by Eastwood Services: 01502 478165; **Insulation** Warmcel cellulose fibre; **Wind Turbine** Proven Engineering, Kilmarnock: 01563 543020; **Solar Collector** AES, Forres: 01309 676911; **PV Array** Wind and Sun 01568 760671; **Condensing Boiler** Eco-hometec: 01302 722266; **Thermal Store** Acu-tank, ACV: 01383 820100; **System design & installation** EUFS, Glenleg, 01599 522288.

BREAD

# BRICKS AND LIME MORTAR

**Colin and Margaret Williams' new self-built home combines energy-saving features with practical style.**

WORDS: CLIVE FEWINS PHOTOGRAPHY: JEREMY PHILLIPS

At first sight, Colin and Margaret Williams' house, with its neat brickwork and interlocking dark clay-tiled roof, looks just like the bordering houses in the quiet road on the edge of a village near Norwich. Take a closer look, however, and you realise there is something different. The bricks are a lot lighter in colour than those in the surrounding houses and are laid in a different way, the sitting room at the front has an unusual glazed angular frontage, and the timber and aluminum coupled sash windows are of an unusual design for a British home.

'Lime House', as Colin and Margaret decided to call it when they started the build in 1997, is in fact the winner of the £5,000 first prize in the Norwich and Peterborough Building Society's 2003 Eco Self-build competition. While Colin and Margaret hesitate to label their new home an 'eco house', it contains a large number of ecological features for a house of only 136m$^2$.

You almost walk over the first of these when you approach the front door. An inspection cover conceals a large underground tank for the

**The unusual staircase was designed to make the most of the double height entrance hall.**

**The heavy Norfolk floor tiles and brick internal walls help store heat from the sun during the day, reducing heat demand at night.**

rainwater recovery system. The rainwater is filtered and distributed around the house by a pump that is triggered whenever a toilet is flushed or the tap at the utility room sink is turned on. The water is also used in the washing machine and the outside garden tap. After two years of living in the house, Colin and Margaret reckon the system has reduced their annual water consumption to just under half of what they would normally expect to draw from the mains.

Enter the house through the 2.3m high double timber front doors and you are immediately confronted by the curved steel staircase. "This is not really an eco feature, but what we wanted," says Colin, who designed the house and built well over half of it himself. "We wanted a staircase that was 'different' – that made the most of the double height space we had allocated for it and did not look 'fitted' in the sense of a conventional wooden staircase."

'Different' is a good word to describe the downstairs interior of the house. Although there is a door between the hallway and the kitchen, the rest is open plan. The kitchen floor is striking. It is of Norfolk pamments – terracotta floor tiles – made by a one-man company in a village a few miles away. "We were keen to support local producers and suppliers wherever possible," explains Colin.

The house is run on just 1,300 litres of oil – approximately one tanker load – providing cooking and heating for an entire year. "If we were heating the whole house with radiators I estimate we would need twice as much fuel." says Colin. He goes on to explain that thermal mass plays a vital role in this because the heavy pamment floor and masonry walls – many of the rooms have exposed brick walls – help to even out temperature swings and store solar gains during the day to be released in the evening. "We chose a light brick to create an impression of greater space in what is a fairly deep house that also occupies almost the entire width of the plot," he says.

All the hot water for the heating system comes from the Esse storage cooker. Water warmed by the surplus heat from the range is stored in two large tanks situated in a concrete block-walled room off the kitchen which in itself acts as a thermal store. The radiator water is stored in the twin-walled larger tank and the smaller one is a conventional tank for storing hot and cold tap water. "Really the system is based on adding a thermal store to use surplus heat and not running the storage cooker continuously," says Colin. "We are extremely pleased with its performance. Our only other form of heating is a 5Kw log burning stove in the sitting room."

The Williams' sitting room is at the front of the house and approached up two steps, placed there in order to separate this area from the kitchen. It has a European redwood floor and ceiling – again to indicate that it is a different 'living space' from the kitchen, with its clay floor and plaster ceiling. The sharply-angled frontage – just three large panes of glass – has been designed to make the most of the morning sun.

The exterior walls are all solid brickwork. After completing his first family

"ALL THE HOT WATER FOR THE HEATING SYSTEM COMES FROM THE ESSE STORAGE COOKER"

self-build, a cavity-walled house on the adjoining plot – Lime House is built in what was its garden – Colin decided solid walls were far better.

"There are many things wrong with cavity walls," states Colin. "In my view the main problems centre on poor site practice, making it almost impossible in some circumstances to stop excess mortar bridging the cavity. These types of wall are also leaky – the average cavity walled house loses about half its heat through all the little gaps – and when bricks are used on the outside it is a dishonest form of construction."

Colin is referring here to the fact that 215mm (9") solid brickwork allows the use of a proper brick bond. "With cavity walls, in which the brick outer skin is non load-bearing, we see mile upon mile of boring English bond," says Colin. "When the builders have decided to do something more exciting they have to resort to breaking bricks ('snap headers') to make them look like something else – the headers in a traditional Flemish bond like we have used here, or a garden wall bond."

**An oil-fired Esse storage cooker provides all the hot water for the heating system.**

Colin has used a British hydraulic lime mortar throughout. "It looks good, has far more stability than a Portland cement mortar, and it allows water entering the masonry to evaporate out," he explains. "Unlike Portland cement mortars, lime also allows the bricks to be reclaimed when the

"THE COUPLED SASH WINDOW SYSTEM INCORPORATES TWO PANES OF GLASS WITH A SPACE OF 70MM BETWEEN TO CONTAIN BLINDS."

building has come to the end of its life." Colin estimates that he laid 60 per cent of the 35,000 bricks (from the Blockley Heritage range), all the pamment floors, more than half the rooftiles and did 20 per cent of the internal, wooden-floated, rough textured lime plaster walls. Apart from the bricklaying, most of the rest of the labour came from his four sons, all of whom have left home but were all able to lend a hand at some stage, and Margaret, who became an expert mixer and hod carrier.

Being a Chartered Building Surveyor and local authority building control officer, Colin is well versed with insulation requirements and was able to meet them by using a 180mm inner timber frame using timber studs, 100mm Warmcell recycled newspaper insulation, a vapour control membrane and lime plaster over chicken wire reinforcement attached to Heraclith strand board. The latter is made from wood waste bound together with magnesite, a natural cement.

All the windows are from Norfolk-based Sampson Windows. "The coupled sash system incorporates two panes of glass with a space of 70mm between to contain blinds," says Colin. "This system offers better insulation values than sealed units but without the risk of losing the 'seal' as so often happens with the latter. Glass has a relatively short life when used in sealed units. In this style of unit the glass will last indefinitely."

With such a tightly-sealed house some form of ventilation system is needed. Colin and Margaret chose a whole house passive stack Passivent ventilation system. Three separate insulated stacks extract from the two bathrooms and the ground floor utility room and WC. There are vapour-sensitive grills in 'wet' rooms which automatically open and close depending upon levels of humidity. In dry rooms trickle vents are incorporated within the windows. "We have no condensation whatsoever, which is ample proof of the effectiveness of the system," says Margaret.

Because the family made up about 75 per cent of the labour force – saving around £80,000 – Colin and Margaret took nearly four years to complete the build. However, as if that was not enough, the couple, who have six grandchildren, have just completed a matching one-storey complex in the same style and materials in the garden. It will house a workshop, store and a home office for Colin. They have also bought a plot in a nearby village on which they plan to construct their third self-build. This time it is to be of earth construction and they used their £5,000 prize money from the Norwich and Peterborough Award to fund a trip to the southern states of the USA where they studied styles of new earth-built houses. Watch this space...

**Colin and Margaret use a rainwater recycling system which provides water for flushing toilets, the washing machine and garden tap. As a result they use less than half the usual amount from the mains.**

## FACT FILE

**Names:** Colin and Margaret Williams
**Professions:** Building control officer and classroom assistant
**Area:** Norwich
**House type:** Three-bedroom detached
**House size:** 136m²
**Build route:** Self-managed
**Construction:** 215mm solid brick walls with insulated inner timber frame
**Warranty:** Architect's Certificate
**Sap rating:** 100
**Finance:** £85,000 Barclay's self-build loan
**Build time:** May '97-May '02
**Land cost:** Already owned
**Estimated value** £35,000
**Build cost:** £95,000
**Total cost:** £130,000
**House value:** £250,000
**Cost/m²:** £698 (75% own labour)

**48%**
**COST SAVING**

**Cost breakdown:**

| | |
|---|---|
| Substructure | £3,000 |
| Flooring (inc. sub-floors) | £4,100 |
| First floor construction | £3,000 |
| External walls | £16,500 |
| York stone parapet copings, sills and associated leadwork | £1,500 |
| Windows and doors (inc. labour) | £13,000 |
| Internal linings | £4,000 |
| Doors | £3,500 |
| Roof | £5,000 |
| Steel beams | £500 |
| Leadwork | £900 |
| Roof insulation and ceilings | £2,000 |
| Internal walls | £2,700 |
| Stairs | £2,500 |
| Drains | £1,500 |
| Rainwater systems and drains | £3,200 |
| Kitchen and bathrooms | £6,500 |
| Heating and plumbing | £6,500 |
| Electrics | £3,100 |
| Esse storage cooker | £2,700 |
| Plant hire | £3,000 |
| Decorating materials | £800 |
| Woodburner | £1,000 |
| Miscellaneous | £4,500 |
| **TOTAL** | **£95,000** |

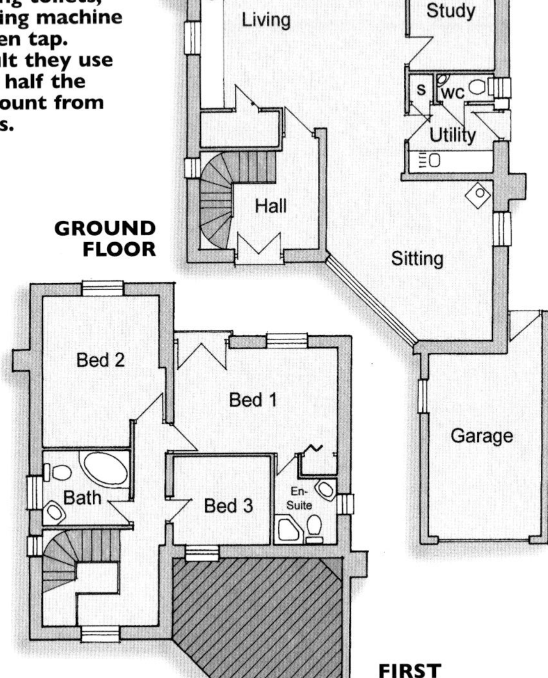


## FLOORPLAN

**A large open plan kitchen/ living room merges with a sitting room downstairs, which has an angular window overlooking the front garden.**

## USEFUL CONTACTS

**Designer** Colin Williams: 01362 850171; **Windows** Sampson: 01449 722922; **HF Roof tiles** J Medler: 01603 264466; **Terracotta floortiles** Norfolk Pamments: 01508 532529; **Staircase** Wensum Engineering: 01328 862703; **Esse Range** Ouzledale Foundry: 01282 813235; **Organic paints and stains** Auro: – Blockleys: 01952 251933; **Hydraulic lime** Hydraulic Lias Limes: 01935 817220; **Woodburning stove** Clearview: 01588 650401; **Rainwater recovery system** Rainwater Harvesting Systems: 01452 772000; **Warmcell insulation, Bitvent and Sarket** Excel Industries: 01495 350655; **Heraclith boards** Thermica: 01482 348771. www.heraclith.com; **Douglas fir kitchen units** Woodwork Shop: 01328 829529; **Galvanised steel gutters and downpipes** Lindab: 01604 707600; **Bespoke doors** David Adcock: 01362 860231

# MAKING WAVES

Robert and Lesley Watson have built a contemporary and environmentally friendly family home with curves in all the right places.

WORDS: DEBBIE JEFFERY PHOTOGRAPHY: NIGEL RIGDEN

Robert and Lesley Watson moved to Devon to live near the sea, so it seems only right that their undulating new house should be constructed in the shape of a wave. The contemporary single storey property is surrounded by mature trees and was built in structural steel and timber frame, with extensive south-facing glass to take advantage of the sea views. It also incorporates a number of eco-friendly features and was designed to accommodate not only the Watsons and their son, Joseph, but also Lesley's mother, Barbara, who lives in her own suite to one end of the three-bedroom house.

"We'd always rather liked the idea of building our own home, and started out by looking for plots in the Worcestershire area," explains Robert. "Eventually we acknowledged that we would really prefer to be by the sea and began hunting for land in Devon instead. The problem was that moving down to the West Country would take us even further away from Lesley's mum, who lived in Yorkshire, so she decided to join us."

Finding land in a totally different region proved difficult, so Robert – a librarian – negotiated to work a three-day week so that he would be able to devote the remainder of his time to plot hunting. "It wasn't a hardship," he smiles, "and without putting in the time I doubt whether we would have heard about this plot."

After three months of searching, the couple saw an advert for a sheltered garden plot in the traditional seaside town of Seaton, which is

**There are 27 windows – including three sets of French doors – on the south side of the building looking towards the sea. The solar panel, mounted on the overhang, helps to warm the domestic water and contributes towards the underfloor heating.**

"I KNOW THAT PEOPLE GENERALLY RATE EN SUITES AS HIGHLY DESIRABLE, BUT WE THINK THEY ARE MORE SUITED TO HOTELS THAN HOMES"

situated in an Area of Outstanding Natural Beauty. Covering one third of an acre, the site benefits from sea views and boasts a number of established plants with a small stream running across the land. Outline Planning Permission had already been granted for a conventional bungalow, but Lesley and Robert had rather different ideas.

Despite the fact that they had always lived in traditional houses, the couple hoped to build something altogether more contemporary, which would incorporate environmentally friendly materials and technologies. They spoke to the planners to determine whether a modern eco house would be viewed favourably, and were pleasantly surprised by the response. "They told us it was just what Seaton needed, and were on our side from the start," says Lesley.

The Watsons contacted a local architectural practice to discuss their specific needs, and were determined that the design of their new home should be led by the shape of the plot. Planning restrictions dictated that the building would need to be single storey and without dormers, which resulted in a low, wide design with a wall of glazing facing south towards the sea views and maximising passive solar gain.

"I don't remember who first suggested that the house should be curved, but we loved the idea that the wave form would create curved walls and irregularly shaped rooms," Robert remarks. "Of course, the organic shape did make it a little more difficult to build because builders don't like curves. We also chose to have a slightly sloping planted roof to allow the house to blend in more easily and keep a low profile on the site."

The layout was designed so that Barbara had her own suite of rooms, which benefit from a separate entrance lobby inside the main front door and open onto a west-facing patio space. Robert and Lesley's accommodation is predominantly open plan, with a spacious living/kitchen/dining space facing south onto a second patio. This area opens into a separate study, which doubles as a guest bedroom, and there are two further bedrooms, a bathroom, shower room and walk-in wardrobe located to the east end of the building. "We decided against en suites and chose to have bathrooms which are reached from the main hallway rather than through the bedrooms," Lesley explains. "I know that people generally rate en suites as highly desirable, but we think they are more suited to hotels than homes."

The north side of the house has minimal glazing to reduce heat loss, and these long, narrow windows have been echoed internally with six similar slot-shaped openings in the partition walls between the south-facing living rooms and the curving internal hallway. Not only do these allow natural light to pass through into the darker part of the building, they also proved a practical alternative to fitting expensive rooflights or sunpipes into the green roof. The slots form unusual focal points that link the different areas of the house and may be illuminated from above by uplighters.

Glulam beams enable wide, open plan spaces within the house, and are supported on a structure of exposed steel posts. The kitchen, dining and living areas are one large, open plan space. The free-form wave design of the house is echoed internally with curving walls and fittings, while internal slot openings bring light into the curved hallway from the south-facing rooms and visually connect the spaces.

**Lesley and Robert built the colourful kitchen area using bamboo worktops, a Corian sink and pale blue cabinets from IKEA.**

"PHOTOVOLTAICS WOULD HAVE BLOWN OUR BUDGET AT THE TIME AND HAVE SUCH A LONG PAYBACK PERIOD THAT THEY WEREN'T ECONOMICALLY VIABLE"

Planning permission for the unusual design was swiftly granted, and Robert and Lesley moved onto the site to live in a two-berth caravan for six months during the build. There was, however, a brief period when Lesley – unable to stand caravan life for a moment longer – moved into an apartment for a few days of respite from the appalling wet weather.

As project manager, Robert found living on site invaluable and was on hand to tackle a number of tasks once the structural steelwork and timber framed skeleton had been completed. He insulated the building using environmentally friendly sheep's wool insulation, laid the underfloor heating and dammed the stream to create a natural pond in the woodland garden.

"By this time I had retired from work completely and could dedicate my time fully to building the house," explains Robert, who has recently created the natural-looking Grassgrid driveway and planted 1,800 sedum plugs on the roof deck. "I particularly chose our builder because he was happy to work alongside me on site, and didn't mind that I wanted to get my hands dirty."

Lesley teaches interior design and was determined that, despite the property's Modernist white facade, the interiors would include a variety of mood-enhancing colours using eco-friendly paints. "Colour is often shunned in favour of white," she says. "I wanted to show that you don't have to live in a blank canvas and, here, every room has a distinct colour relating to its use."

The master bedroom is painted a restful lilac and enjoys wonderful sea views, while the bathrooms are a more vibrant turquoise. Baby blue kitchen cabinets stand against an orange feature wall, with a creamy Corian sink that looks like white chocolate, and bamboo worktops cut to fit the curve of the room. Even the multicoloured voiles at the windows cast a rainbow of shadows when the sunlight hits them.

Externally, the windows are finished in an olive green colour and the glass is shaded by a slate-covered overhanging section of roof, which prevents the house from overheating during the summer. A solar panel has been mounted on this overhang which helps to warm the domestic water and contributes to the underfloor heating. The Watsons also hope to add photovoltaic solar panels in the future which will generate electricity.

"Photovoltaics would have blown our budget at the time and have such a long payback period that they weren't economically viable," remarks Robert. "We wanted to stick to our original £200,000 budget, and funded the build from our savings rather than a mortgage. Building with curves did bump up the cost, but we managed to find ways of cutting back. The plot also cost more than we originally planned, but it's in such a wonderful position that we had to have it, and we thoroughly enjoying living so close to the sea in such a beautiful new home."

"I DON'T REMEMBER WHO FIRST SUGGESTED THAT THE HOUSE SHOULD BE CURVED, BUT WE LOVED THE IDEA THAT THE WAVE FORM WOULD CREATE CURVED WALLS AND IRREGULARLY SHAPED ROOMS"

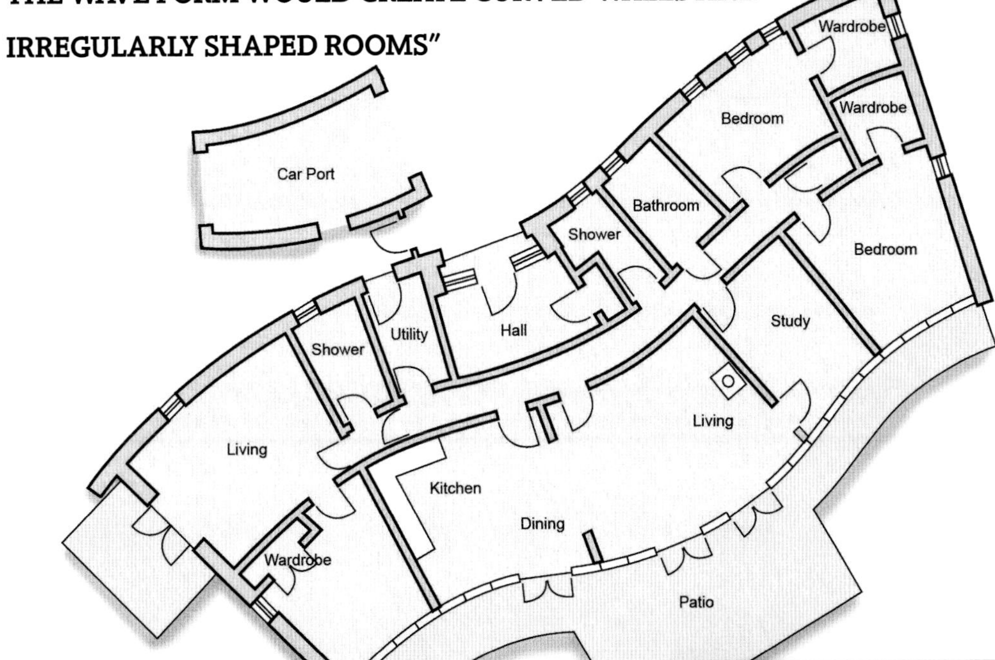


## FLOORPLAN

**Most of the main living rooms have been placed to the south of the plan, which enjoys direct sunlight and sea views, while bedrooms and bathrooms are positioned to the north. The living room is open plan to the kitchen and dining areas, and Lesley's mother has her own suite with a bedroom, bathroom and lounge.**

## FACT FILE

**Names:** Robert and Lesley Watson
**Professions:** Librarian and interior designer/teacher
**Area:** Devon
**House type:** Three/four-bedroom single-storey house
**House size:** 200m$^2$
**Build route:** Builder and specialist subcontractors
**Finance:** Private
**Construction:** Timber frame and planted roof
**Warranty:** Architect's certificate
**Build time:** Jan '04-June '05
**Land cost:** £150,000
**Build cost:** £205,000
**Total cost:** £355,000
**Current value:** £450,000
**Cost/m$^2$:** £1,025

**21%**
**COST SAVING**

**Cost breakdown:**

| | |
|---|---|
| Service connection | £2,379 |
| Groundworks | £21,134 |
| Structure | £110,431 |
| Windows and doors | £23,324 |
| Plumbing and heating | £20,544 |
| Electrics and lighting | £8,312 |
| Green roof | £16,455 |
| Solar panel | £1,722 |
| Fees etc | £13,249 |
| VAT refund | £-12,507 |
| **TOTAL** | **£205,043** |

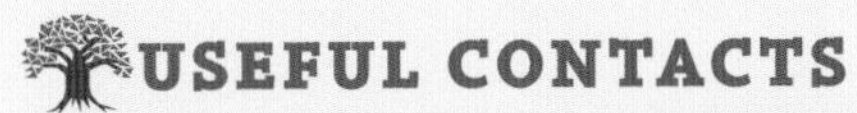

## USEFUL CONTACTS

**Architect** ARA Architect: 01395 271619 **Builders** A & J Carpenters: 01395 519300 **Green roof components** Blackdown Horticultural Consultants Ltd: 01460 234582 **Grassgrid driveway** Charcon: 01335 372222 **Eco paints** Ecos Organic Paints: 01524 852371 **Thermafleece sheep's wool insulation** Second Nature: 01768 486285 **Windows and French doors** Rationel Windows: 01869 248181 **Solar panel** Wessex Solar Systems: 01202 517526 **Eco products** The Green Shop: 01452 770629 **Underfloor heating** Nu-Heat UK Ltd: 0800 731 1976 **Kitchen units** IKEA: 020 8208 5600 **Timber** Jewson: 02476 438400 **Cushion flooring** Amtico: 0121 745 0800 **Bamboo worktops** Reeve Flooring: 01485 210754 **Pebble mosaics** Tile Clearing House: 01392 411110 **Sanitaryware** Ideal Standard: 01482 346461 **Sundries** Bradford Timber: 01297 20123 **Driveway** CSD Groundworks: 01395 568005 **External doors** Devon Joinery: 01395 239049 **Prelasti roof construction** DRM Fabrication: 01373 830924 **Woodburner** Exeter Stoves: 01392 410903 **Gas boiler** Ford & Sons: 01395 571000 **Electrics** Gator Electrical: 01395 513149 **Dry-lining** Hugo Headon: 01395 568919 **Drainage components** Mole Avon: 01363 774786 **Structural engineers** Nicholls Basker: 01626 776121 **Corian sink** PB Kitchens: 01297 22559 **Steelwork** Parkins Engineering: 01395 233368 **Plumbing** PIW Services: 01395 268227 **Scaffolding** Sidmouth Scaffolding: 01395 515896 **Gravel, pebbles and concrete** Westcrete: 01297 32002

Andrew and Louise James have built an attractive chalet-style home on a modest budget which incorporates many energy-saving features.

# ECO-MONEY

WORDS: HEATHER DIXON PHOTOGRAPHY: JEREMY PHILLIPS

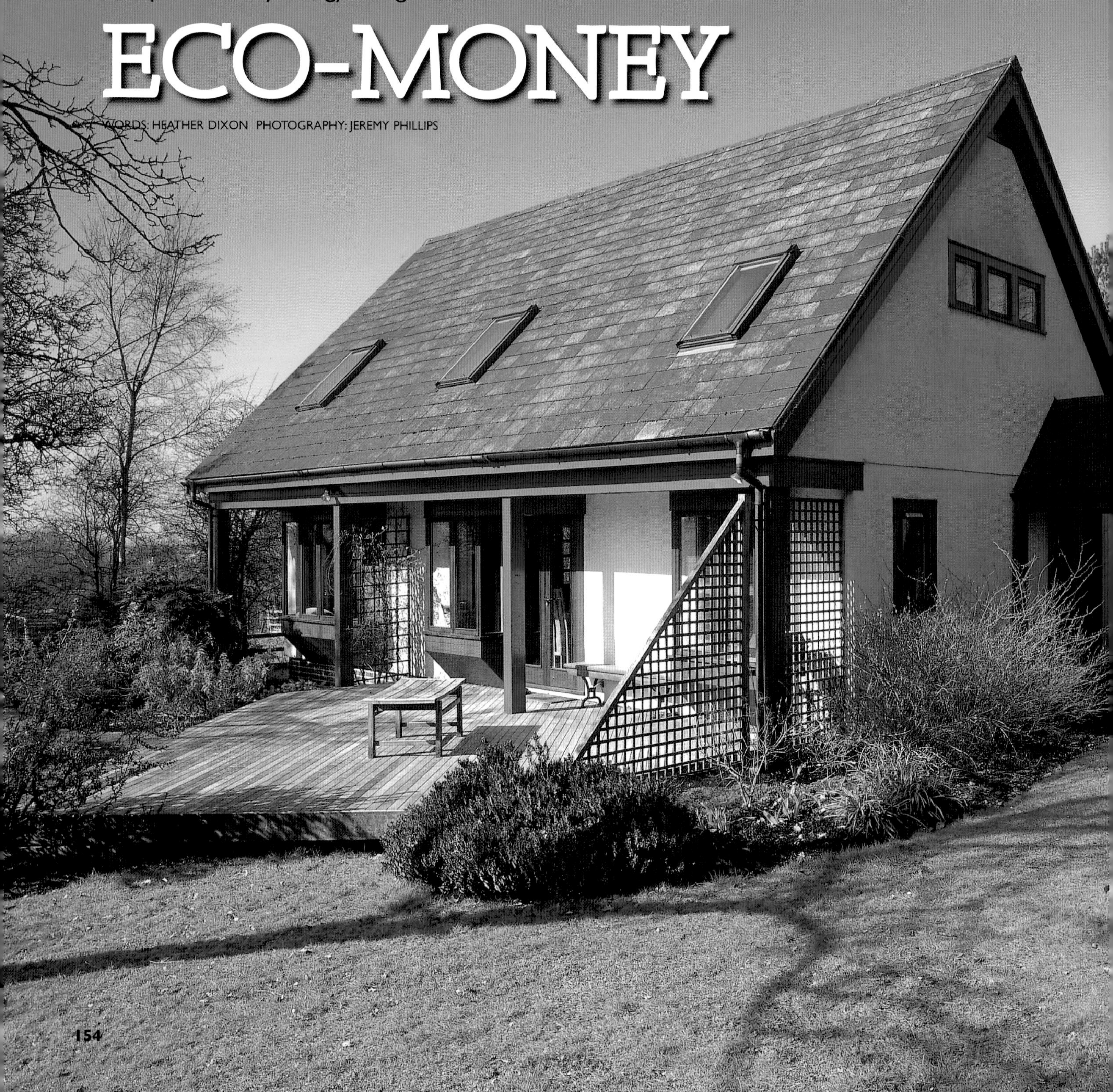

Architect Andrew James was ahead of his time when he started planning a house which would be eco-friendly and open plan. "We came back to Britain from Vancouver in Canada where energy-efficient homes and spacious designs were commonplace," says Andrew. "Over there, temperatures can plummet to minus 40°C so, even as a young man, I was going to seminars about passive solar awareness in relation to the home."

Determined to build a family house which would accommodate as much as possible of what he'd learned, Andrew and his wife Louise started looking for the perfect plot. "It had to be south facing, gently sloping, full of trees and close to a decent pub," he says.

They found the ideal location in a village orchard near Easingwold in Yorkshire, which had permission for a house approved in principle by Hambleton District Council. The only stipulation was that the dwelling had to be built close to the road in keeping with other buildings in the village.

Andrew started work designing a low-energy, space-efficient chalet-style property with rooms in the roof and low overhanging eaves. Following the basic principles of passive solar design, windows were optimised on the south-facing elevation to maximise solar gain, and kept to a minimum on the north elevation where they would lose most energy.

**All the doors, architraves and furniture in the dining room and kitchen were made by Andrew, who started cabinet making as a schoolboy and, even though he trained as an architect, spent many years running his own furniture-making business.**

By the time Andrew and Louise had managed to gain planning and building regulations consent and were ready to start building, 18 months had passed. They had moved into a caravan on site, a cost effective option that also proved highly practical, as Andrew planned to do much of the building work himself in order to keep down costs, including digging the foundations and laying the foundation slab which he and the builder did between them.

Andrew chose to build using a timber frame rather than the more conventional brick and blockwork. As a skilled carpenter and furniture maker, using timber seemed like a perfectly logical choice. The frame he chose is of standard 89mm stud depth, filled with mineral fibre insulation. Inside this, however, Andrew has created a second insulating layer using foil faced, polythene bubble-wrap insulation fixed to the inside of the frame with spacer battens behind the internal plasterboard finish. The battens used are sufficiently deep to leave a clear airspace behind the plasterboard, both as a convenient service duct for plumbing and wiring, and also to allow the reflective foil to function – the foil only reflects radiant heat if there is a void in front of it. The joints in the polythene insulation are taped together using reflective foil tape to create an airtight seal and vapour control layer.

**The fireplace, created with tiles on plasterboarding, is fitted with a woodburning stove which has a steel sheet bolted to the back to reflect the heat. All the wood used for the fire comes from windfalls and dead wood in the garden.**

"The thermal performance of the house is to a specification which still

## "THE THERMAL PERFORMANCE OF THE HOUSE IS TO A SPECIFICATION WHICH STILL EXCEEDS THE LATEST BUILDING REGULATIONS"

exceeds the latest building regulations," says Andrew. "The whole thing about designing and building energy-efficient houses is that the air tightness and fitting of the insulation is absolutely critical. If you are not a perfectionist, the whole principle is wasted because it will leak air all over the place and, along with it, the heat. That was the reason we wanted a timber-framed house. It was easier to make airtight and insulate to the standard we required."

Effectively, the house works like a large, airtight survival blanket. The rate of air change, other than from the opening and closing of external doors and windows, is controlled using a mechanical assisted ventilation system. This extracts damp stale air from the bathroom, toilet, kitchen and utility room, with incoming fresh air supplied through trickle vents in the habitable rooms. Hot water heating is supplied to an underfloor system by a Tri-Save Turbo 45 condensing boiler and the house, which measures around 150m$^2$, has a heating cost which has averaged less than £300 a year.

All of the windows, apart from the Velux rooflights upstairs, were designed by Andrew and made in Douglas fir by a local joiner. They incorporate shoot bolts and stainless steel friction stray hinges. The Douglas fir doors open onto the red cedar decked outdoor area, and the front door was made in Andrew's workshop. "The porch is a significant part of the house because I wanted to design it in a way which drew people in, but maintained a link with the outside," says Andrew.

Outside, the walls are rendered. Andrew applied the render onto galvanized steel mesh, fixed to pressure-treated timber battens to create a vented cavity, over a breather membrane on the plywood sheathing.

The softwood roof timbers were cut on site, insulation is installed at ceiling level, in conjunction with the same foil faced bubble wrap system as was used in the walls. The roof areas are floored to provide large, easily accessible storage space. Unusually for this country, the roof has a 1.5 metre overhang on the south side, giving protection from the intensity of summer sunlight and avoiding the risk of overheating while maximising the solar gain and light from the low angle winter sun. "Light is very important to me," says Andrew. "When you are working with a very small volume you can use natural daylight to draw the eye from a small space into the adjacent space to make the overall area seem bigger. The lighting has been a huge success in this house – it varies with the seasons and transforms the interior in different ways all year round. This house is everything I could have wanted – and more."

## FACT FILE

**Names:** Louise and Andrew James
**Professions:** Lecturer and architect
**Area:** North Yorkshire
**House type:** Three-bed detached timber frame
**House size:** 150m$^2$
**Build route:** Self and contractor
**Construction:** Timber frame
**Finance:** Halifax mortgage
**Build time:** Ten months
**Land cost:** £16,500
**Build cost:** £60,000
**Total cost:** £76,500
**House value:** £350,000
**Cost/m²:** £400

**79%**
**COST SAVING**

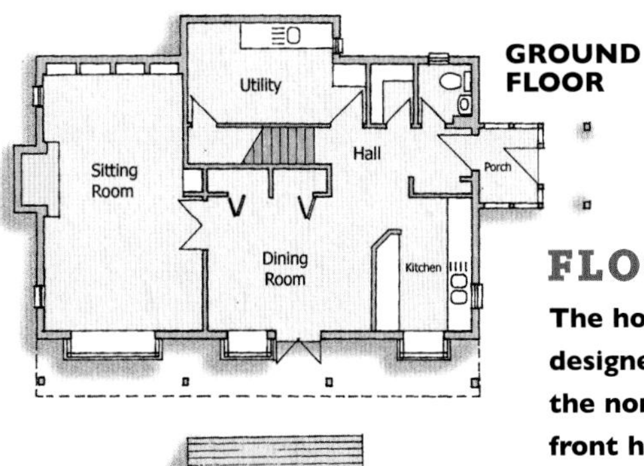


**GROUND FLOOR**

## FLOORPLAN

**The house has been designed so that the north-facing front has minimal fenestration, while the southern elevation has large areas of glazing to maximise passive solar gain.**

**FIRST FLOOR**

## USEFUL CONTACTS

**Architect** Andrew James: amj@designworks2.co.uk 01360 771810; **Joinery** Richard Thompson at Tollerton: 01347 838387; **Oak flooring** Junckers: 01376 534700; **Heating** JSD: 01609 748776; **Slate Roofing** Eternit Slate: 01763 260421; **Foil-faced Insulation** Alreflex 2L2 from Thermal Economics Ltd: 01582 450814.

# BUILDING AN AMERICAN-STYLE TIMBER-FRAME HOME £380,000

**The site is wooded and slopes downwards to the west, affording glorious views and sunsets – features that the design makes the most of.**

WORDS: MARK BRINKLEY
PHOTOGRAPHY: ANDREW LEE

# AMERICAN REVOLUTION

**Peter and Veronica Burbridge have built a new home using an American-inspired construction system.**

Peter and Veronica Burbridge have spent much of their adult lives living in a large but somewhat dilapidated house in Perthshire. They had long toyed with the idea of building a new house in the grounds and as long ago as 1989 had got as far as gaining outline planning permission. At the time, though, the idea of building didn't exactly inspire them, and they failed to renew the consent, eventually allowing it to lapse.

The old house took a lot of maintenance, however, and when the roof started leaking one mid-winter night, the couple finally came to the conclusion that it was time to move on.

"Veronica said to me that we had to either get on and build the house or move somewhere else," says Peter, who had spent much of his childhood in America and nursed a desire to create a North American-style timber house. In 1998, therefore, he flew to the States and visited a number of timber frame homebuilders. They were universally enthusiastic about his project and were all prepared to export one of their homes across the Atlantic to Scotland – but there was the small matter of planning permission to be sorted and Peter realised that for this purpose it would probably be more effective to work with British businesses.

Initially, Peter asked a firm of local architects to draw up some plans. "This process took several months and turned into an incredible waste of time and was, in retrospect, an expensive mistake," says Peter. "I kept telling them what I wanted and they kept re-interpreting it in ways neither Veronica nor I liked."

All the time, though, Peter was getting a stronger feel for what he really wanted and on another trip to the States he saw a house which he thought would be perfect for them. On returning to Scotland, he contacted another architect, John Freeman, whom he and Veronica

**Peter and Veronica employed oak specialists TJ Crump Oakwrights to build the frame. They were more than happy to come up with a post and beam system based on a softwood, Douglas fir.**

"ON ANOTHER TRIP TO THE STATES, PETER SAW A HOUSE WHICH HE THOUGHT WOULD BE PERFECT FOR THEM"

had known for many years. "I explained to John how the design process had come off the rails and I was able to show him photos of what we really wanted to achieve. John was able to draw these up and suddenly we had a design that we wanted to build. It went through the planning process with a minimum of fuss."

Peter and Veronica then sold their old home and moved into a rented house in nearby Comrie. Before they could start the build, they had to work out the access arrangements for the new plot and this was to prove problematic. They had originally envisaged the site being serviced from a small lane at the bottom end of the garden but they couldn't establish who owned the lane. After losing several months in futile negotiations with neighbouring landowners, they decided to cut their losses and change the plot access around completely so that they would be able to approach the house from the high ground above, next to their old house. "It's arguably not such a good way to approach the house but we were keen to press ahead and it seemed a reasonable compromise to reach," says Peter.

**Thanks to the high levels of insulation provided by the SIPs system, there is no need for a conventional central heating system in the house. Instead, the Burbridges rely on an Aga and wood burning stove.**

The next stage in the process was to decide how to get the house built. Peter had seen Structural Insulated Panel (SIPs) houses being built in the States and was keen to use something similar. By now, SIPs had begun to appear in the UK and there were a small number of contractors who were showing an interest in working with them.

Peter interviewed each supplier and selected Hereford-based builders TJ Crump Oakwrights: Peter liked their flexibility and enthusiasm and the fact that founder, Tim Crump, was also a keen student of the American methods of housebuilding. Crump took on the main contract and chose to subcontract the manufacture and installation of the SIPs to BPAC, based in Fife.

Before TJ Crump could get on site, the ground had to be levelled and a large North American-style basement was constructed. The post and beam frame was built in Crumps' yard in Herefordshire, shipped to site and

"WE HAD INTENDED TO BUILD A SMALL CONSERVATORY BUT THE FIRM WENT BUST, TAKING WITH IT OUR DEPOSIT OF £2,500"

erected in just three weeks in May 2001. TJ Crumps mostly work with oak but they bowed to the American preference for softwoods, notably Douglas fir – which is almost as durable and considerably more cost effective. Oak is really only needed when the posts and beams are to remain exposed to the elements but Orchard House was getting a completely weathertight skin, thanks to the SIPs which were craned into place around the walls and on the roof.

The finishing of the house was supervised by Peter, who leaned heavily on the skills and contacts of Robin Spearing, a master carpenter who travels widely across the Highlands supervising the erection of timber frames. The house has chestnut flooring throughout on both floors and a wrap-around balcony on the southwest corner where the views are best. "We had intended to build a small conservatory there," says Peter, "but the firm went bust, taking with it our deposit of £2,500. In fact it turns out to be a lovely space just as it is, so we have no intention of ever covering it now."

By British standards, The Orchard is a large house – its size accentuated by the use of cathedral ceilings upstairs, which makes for spectacular bedrooms and bathrooms. Because of the high levels of insulation afforded by the SIPs, it doesn't require a conventional space heating system. The bulk of the heat is supplied by an Aga and a wood burning stove, both of which were previously in the Burbridges' old house. The bathrooms have electric underfloor heating mats and elsewhere heat is provided by an air-to-air heat exchanger. A Genvex air to water heat exchanger, located in the basement, supplies hot water.

The aluminium-clad timber windows and doors, made by Canadian firm Loewen, were desirable both for their quality and looks. The walls have been finished externally with weatherboarding in a mix of Western Red Cedar and native larch.

The resulting home brings a touch of the United States to Scotland, and is a clever combination of traditional English craftsmanship and modern American construction technology. Peter's determination to build the home of his dreams will pay dividends for years to come.

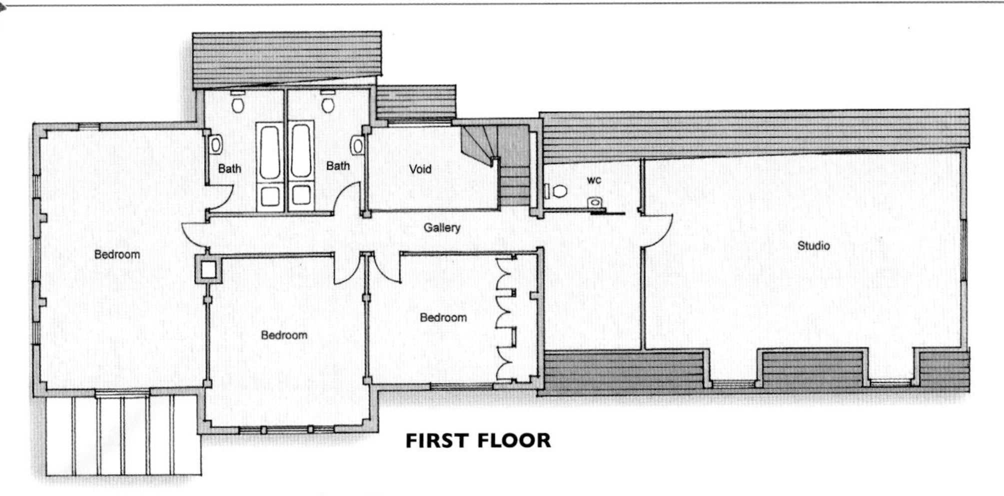


FIRST FLOOR

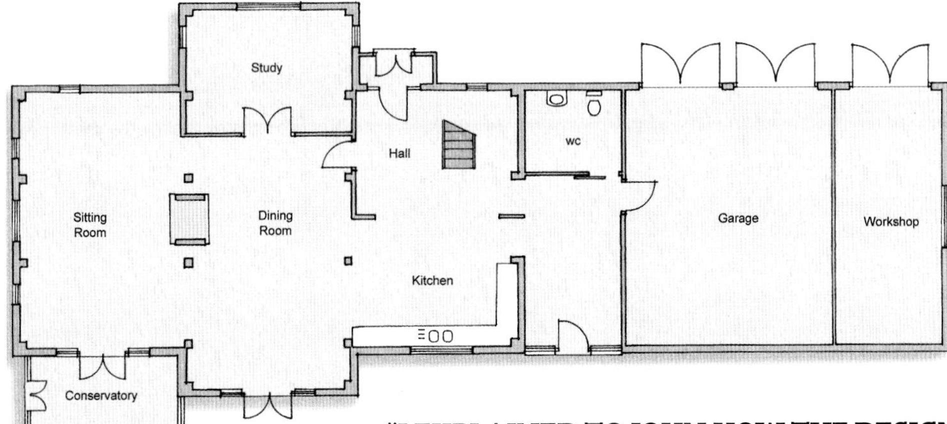


GROUND FLOOR

## FLOORPLAN

**An open-plan sitting/dining/kitchen room is at the heart of the downstairs layout, with three bedrooms and bathrooms upstairs.**

"I EXPLAINED TO JOHN HOW THE DESIGN PROCESS HAD COME OFF THE RAILS AND I WAS ABLE TO SHOW HIM PHOTOS OF WHAT WE REALLY WANTED TO ACHIEVE. HE DREW THESE UP AND SUDDENLY WE HAD A DESIGN THAT WE WANTED TO BUILD"

## USEFUL CONTACTS

**Timber frame** TJ Crump Oakwrights: 01432 353353; **SIPs** BPAC: 01592 882202; **Windows** Loewen: 01908 310798; **Master carpenter** Robin Spearing: 01764 685245; **Architect** – John Freeman: 01764 653345; **Electricians** J Halley: 01764 652814; **Mason** J Jack: 01764 654403; **Plumber** S Roberts: 01764 679521

## FACT FILE

**Names:** Peter and Veronica Burbridge
**Professions:** Environmental consultant and parliamentary liason officer
**Area:** Perthshire
**House type:** Four-bedroom with basement and garage block
**House size:** 360m² + 108m² basement + 57m² garage
**Build route:** Builder to weathertight shell, managed subcontractors to finish
**Construction:** Douglas Fir post and beam with SIPs
**Finance:** Private
**Build time:** 12 months
**Land cost:** already owned
**Build cost:** £380,000
**Total cost:** £380,000
**House value:** £500,000 est.
**Cost/m²:** £800 inc. basement

**24%**
**COST SAVING**

## ENVIRONMENTALLY FRIENDY BARN CONVERSION £300,000

The barn was originally built in 1840 by the Rothwell Estate; reclaimed, natural materials have been used wherever possible to preserve the character of the building. Insulation exceeds Building Regulation requirements and includes 110mm-thick sprayed insulation to all roofs.

Richard and Carole Lea have converted a small brick barn into a luxurious and environmentally friendly new home.

# A GREEN CONVERT

WORDS: DEBBIE JEFFERY PHOTOGRAPHY: JEREMY PHILLIPS

At just 150m², Rothwell Barn may be relatively small but it's certainly perfectly formed. The detached brick building literally bristles with a host of technologies which are more often reserved for big-budget projects, and yet this modest conversion was completed for the realistic and affordable sum of £300,000. Even better, the solar panels, rainwater harvesting and geothermal heat pump which Richard and Carole Lea have installed will all dramatically reduce their home's running costs in the future.

"We wanted to produce a home from a redundant 1840s brick barn – one of our farm buildings – so that we could still live on site but could sell the rest of the farm and take things a bit easier in our retirement," Richard explains. "We have four children and seven grandchildren and needed enough space to entertain the family, as well as somewhere which would be comfortable, convenient, secure and cheap to run."

The Leas intended to create authentic features using natural materials — reclaiming as much as possible from the original barn in order to preserve its character. With plans to spend some of their retirement travelling, they wanted their home to have numerous security devices, which could be operated remotely while they were living abroad. It was also important that the converted barn should cost little to run in terms of energy bills, water and sewerage, using the latest technology for saving both money and energy.

"I spent six months researching the equipment I chose and another nine months managing the installation," says Richard. "Carole and I drew up our own floorplans before engaging an architect to help us through the planning stages, and we designed a very simple home with a traditional layout. The ➤

## "IT'S TAKEN A GREAT DEAL OF CAREFUL RESEARCH, PLANNING AND ORGANISATION TO PREVENT A MONUMENTAL BLUNDER..."

front door opens directly into the kitchen, doing away with the need for a space-wasting hallway. There's a central dining area and living room on the ground floor, with four bedrooms, an en suite and bathroom upstairs."

In fact, the couple would have liked to increase the size of their bedrooms by building across above the attached garage, but were told that planning permission would not be granted for this extension to the original building. They were, therefore, both surprised and pleased when they were allowed to add on a sunroom leading off from the main lounge, which creates an additional reception room on the ground floor.

The Lancashire barn might appear simple from the outside, but its massively insulated shell conceals a host of modern and eco technologies which caused the various subcontractors much head-scratching. "I had two main problems along the way," explains Richard, who project managed the build himself.

"My groundworks crew were awful and I had to replace them. Then the first plumber I recruited – who I'd used for a number of years – simply wasn't up to the task, so I had to bring in a more experienced tradesman to deal with the geothermal heat pump, the rainwater harvesting and solar panels. Other than that, installing the equipment has been trouble free and it all works a treat."

One of the first things that Richard decided to invest in was geothermal underfloor heating. "I spent four months investigating the different options and was still totally confused," he recalls, "so I decided to go and visit various suppliers, and this led to choosing coils of pipe, laid horizontally in two-metre-deep trenches, which were easy and comparatively cheap to install."

Instead of also burdening the heat pump with heating their hot water, the Leas decided to install four solar panels, each with ten vacuum tubes (OPC collectors), on both the east- and west-facing roofs, which are supplemented by off-peak electricity immersion heaters in the winter. Bills are further reduced by ensuring that every high user of electrical power in the barn, such as the washing machine, dishwasher and tumble drier, are on an off-peak 'Economy 10' tariff, when electricity costs one third of the usual price for ten hours each day.

The performance and reliability of the various eco systems has proved far better than Richard had originally hoped, and the couple expect to save around £2,000 per year on their energy and water bills, as well as adding £35,000 to the overall value of the house, according to estimates by local ➤

**The windows, external doors and stairs are made from solid oak, with distressed oak internal doors, skirting and flooring in the sitting and dining rooms. The internal exposed brick walls are built from handmade kiln-dried bricks and the stone floors are fossil-rich sandstone from India.**

In addition to a wealth of eco features, Richard has included an audio/video system that accesses a central server from wall-mounted controls throughout the house.

"I HAD TO BRING IN A MORE EXPERIENCED TRADESMAN TO DEAL WITH THE GEOTHERMAL HEAT PUMP, THE RAINWATER HARVESTING AND SOLAR PANELS"

estate agents. "Our rainwater harvesting system takes all the water from the roof through five downpipes and 100mm underground pipes. Once filtered, it's stored in a 4,000-litre underground GRP tank ready for use for toilets, outside taps, the washing machine and fire sprinklers," explains Richard, who was adamant that he wanted sprinklers throughout the ground floor of the barn, despite the fact that finding a company to install the system using a rain-harvesting supply proved difficult.

When the same problem arose sourcing a professional to fit the centralised ventilation and air conditioning systems, Richard decided to give it a go himself, installing them with the help of some friends.

"It seems to me that, when I was a boy, the outside air quality was appalling and often full of industrial smog," he says. "Now, our over-heated, well-sealed homes mean that the air quality indoors is much poorer, so I installed a system which changes the air in the house every two hours – taking it from wet areas such as the bathrooms and kitchen, and extracting 80% of the heat to warm fresh filtered air brought in from outside."

His love of gadgets also led Richard to specify a programmable lighting system that uses mainly low-voltage lighting through a centralised system of dimmers, which is atmospheric, convenient and eco friendly. A combination of LEDs, low-voltage, fluorescent and sodium lights have been used inside and outside the house, with up to five zones of lighting and eight pre-set scenes in each room or garden area.

This lighting system also controls the electric curtains, electronic gates and garage doors, which may be operated from the car when the Leas return home. Additionally, an away-from-home mode draws and opens curtains and switches lights on and off to simulate occupation.

There's also an audio/video system which distributes FM radio, music from a media server, Sky+HD, DVD and CCTV images all through the house to in-ceiling hi-fi speakers in eight rooms and plasma screens in six, as well as to a home cinema in the sitting room – where a large pull-down screen, surround-sound speakers and an HD digital projector are fitted.

"It's taken a great deal of careful research, planning and organisation to prevent a monumental blunder, because some of the products I wanted to use are still relatively new to the UK market," says Richard.

"Although saving energy was important, comfort and convenience were also main objectives – and that's something which can't be costed. Now we are in a position to sell the farmhouse and move into Rothwell Barn and, after all the hard work, we really can't wait. What could be more satisfying than turning on the heating or hot water and knowing that it's not costing you a penny?"

## FLOORPLAN

**Eight cow stalls once stood in what is now the kitchen, an overhead feed store has become the two back bedrooms, and the cart store is a dining room and master bedroom, while a tool/hay store to the side has become the sitting room with a bedroom and bathroom above. The garage was added in the 1950s, and in the 1970s the building was converted to equestrian use. When Carole and Richard bought the farm back in 1997, they converted the tack room and overhead store into an office and continued to use the stables and garage until the end of 2005.**

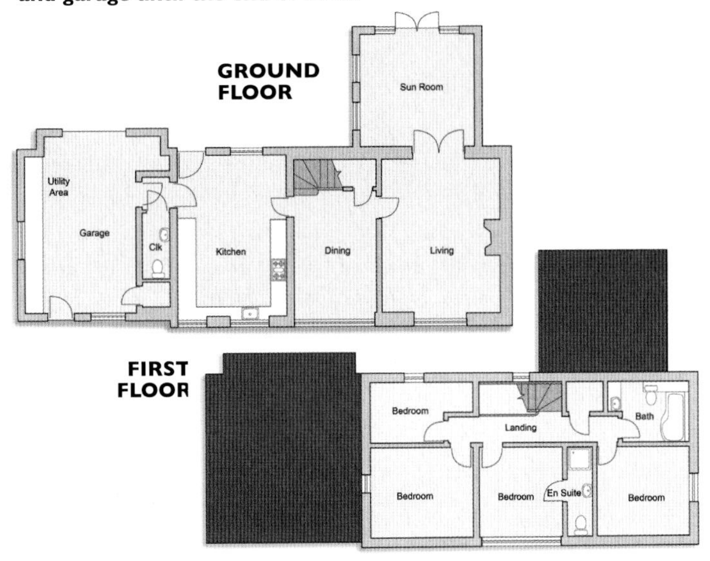


## ECO FEATURES

**● Solar panels ● Rainwater harvesting ● Ground-source heat pump ● Heat-recovery ventilation unit ● Low-voltage lighting system**

## FACT FILE

**Names:** Richard and Carole Lea
**Professions:** Company director and retired hospital manager
**Area:** Lancashire
**House type:** Four-bed barn conversion
**House size:** 152m²
**Build route:** Self-managed subcontractors
**Finance:** Private
**Construction:** Brick and block
**Build time:** Jan-Nov '06
**Barn cost:** already owned, est value £150,000
**Build cost:** £300,000
**Total cost:** £450,000
**Current value:** £550,000
**Cost/m²:** £1,974

**18%**
**COST SAVING**

**Cost breakdown:**

|  |  |
|---|---|
| Ground-source heat pump | £4,653 |
| Heat pump installation | £1,250 |
| Underfloor heating | £2,655 |
| Solar panels | £6,000 |
| Ventilation system | £3,250 |
| Rainwater harvesting | £1,940 |
| Fire sprinklers | £1,950 |
| Lighting system | £7,000 |
| Light fittings/installation | £7,000 |
| Audio/video system | £30,000 |
| Air conditioning system | £3,250 |
| Other building costs | £230,000 |
| **TOTAL** | **£298,948** |

## USEFUL CONTACTS

**General advice and information** Richard Lea: 01772 613179 **Architects, planning and building regulations applications** Cork Toft Partnership Ltd: 01772 749014 **Groundworks** Tony Kirkham: 01772 601843 **Structural building work** Ash Barn Construction Ltd: 01772 690879 **Reclaimed building materials** Martin Edwards: 01772 334868 **Installation of roof insulation** Warmroof: 0870 111 0706 **Patios, fences and drive** Distinctive Driveways: 01282 870945 **Electrical installation** Coxhead Electrical: 01772 751444 **Lighting, audio/video systems design and commissioning** Evolve Total Home Systems: 01524 272400 **Decoration** SL Rogers: 01772 615427 **Plumbing and bathroom installation** ARM Plumbing: 07796 304952 **Advanced plumbing of geothermal, underfloor heating, rainwater harvesting etc** Dutton Plumbing & Heating: 01942 244558 **Sanitaryware and plumbing materials** Mayalls: 01942 241711 **Supply and installation of kitchen, utility room and bedrooms** MFI: 01772 556032 **External doors, windows and stairs** Moor Park Joinery: 01539 561158 **Internal oak doors, skirting and floorboards** Old Time Timber: 01787 277390 **Rainwater harvesting equipment** Rainharvesting Systems Ltd: 01453 836817 **Geothermal equipment** Kensa Heat Pumps: 01326 377627 **Underfloor heating** Continental Underfloor Heating: 0845 108 1204 **Septic tanks** Biodigester Ltd: 01278 786104 **Ventilation equipment** Allergy Plus: 0870 190 0022 **Carpets** Hawardens: 01772 612405 **Trees and plants** Embleys Nurseries: 01772 612227 **Plant hire** Taylor Plant Hire: 01772 614939 **Fire sprinklers** Homesafe Fire Sprinklers: 01706 718404

**The super-insulated timber frame house has been built with low-impact materials wherever possible and is clad in untreated cedar. Sunslates from Solar Century on the roof generate 25% of used electricity, part of which is exported to the National Grid.**

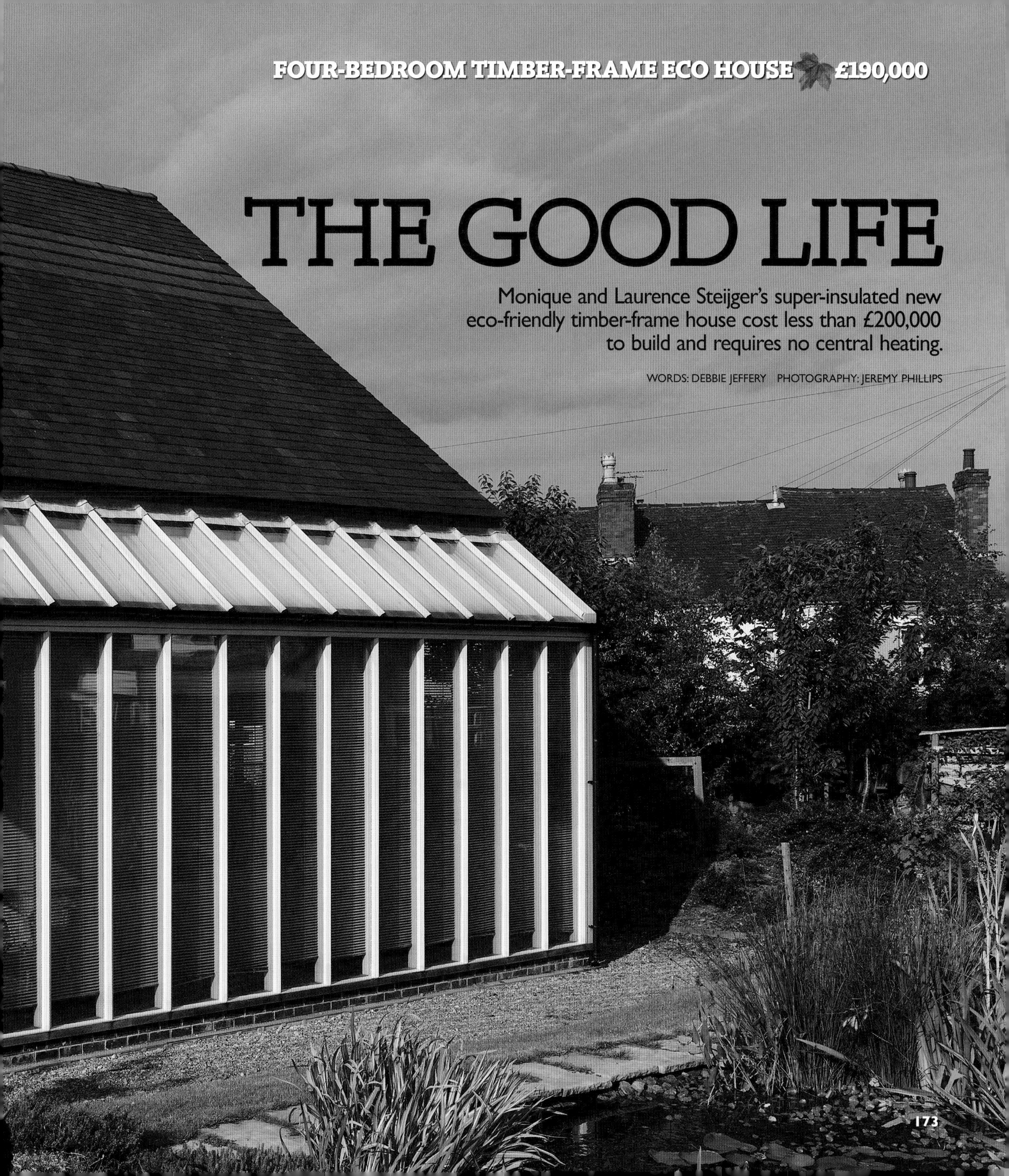

# THE GOOD LIFE

Monique and Laurence Steijger's super-insulated new eco-friendly timber-frame house cost less than £200,000 to build and requires no central heating.

WORDS: DEBBIE JEFFERY PHOTOGRAPHY: JEREMY PHILLIPS

"We built our own home because it was impossible to find an affordable house that fulfilled all our wishes," explains Laurence Steijger. This might sound like the familiar cry of every self-builder, but Laurence and his wife, Monique, had a very specific list of requirements which many people in the UK would find totally alien.

The couple are Dutch, but have lived in numerous countries, including Switzerland, Germany and Sweden – where highly insulated, well-sealed buildings come as standard. Moving to England 10 years ago proved to be something of a culture shock when it came to the standard of housing they rented, and the Steijgers were keen to create a home which would better reflect their green values.

As an energy consultant, it was Laurence who took charge of developing the design and incorporating energy-saving materials and technologies. "Our first thought had been to develop a system of hot-air central heating," he says. "It's a concept that the Romans used, and one that was interesting enough for our architect, Alan Newton, to agree to help us design. When the first quotes came in, though, it became clear that this idea would be unaffordable – so we decided to go for a passive solar house instead."

Building Regulations relating to energy efficiency may have tightened in the UK, but what the Steijgers proposed exceeded anything dreamt of by most building control officers. Their new two-storey detached home was built using a super-insulated timber frame, and incorporates active and passive solar principles. A single high-efficiency wood stove is all that's needed to heat the house. Other than that, the only heat required comes free from the sun's rays. Consequently, their total energy bill is a meagre £240 for an entire year.

**The south-facing conservatory maximises passive solar gain and ensures that the house requires no central heating – just a single woodburning stove for winter.**

In addition, the permanent ventilation system has a heat-recovery unit, which means high air quality, low moisture and low heat loss. Any surplus heat – particularly that taken from the utility room – is used by a heat pump to provide hot water. A 5,000-litre rainwater tank supplies the toilets, washing machine and outside tap, while 15m$^2$ of solar panels create around 1,000 kWh of electricity from daylight each year.

"Finding our plot wasn't easy, because we wanted as many square metres for our money as possible so that we would have enough space to grow food and keep chickens," says Monique, a business analyst. "It was important for us to live within easy reach of main roads and a railway station, but we also wanted somewhere relatively quiet. At last, with our architect's help, we found this building plot in Derbyshire."

The plot in South Normanton is almost 1,000m$^2$, and had previously been used as allotments and for keeping horses. Bordering fields, its peaceful location in the middle of town was perfect. Once cleared by a JCB, the barren-looking plot appeared to double in size. "While we were waiting for

"THE HOUSE CAN BE HEATED BY OPENING THE CONSERVATORY DOORS AND MAKING USE OF PASSIVE SOLAR GAIN"

**Laurence made the kitchen cupboards from plywood frames fitted with solid wooden doors, avoiding chipboard and, therefore, minimising the release of formaldehyde in the house.**

the build to begin, we started working on the garden," says Laurence, who created an organic vegetable plot. "We were picking our first broad beans while the foundations were still being dug."

When it came to the layout of their home, the Steijgers needed to consider a number of factors. Not only did they require four bedrooms and two bathrooms, they also wanted a home office for Laurence's consultancy business. The main living/dining area is open plan to the kitchen, and there are two storage rooms on the north side of the property.

These, and a full-width double-height conservatory on the south side, provide heat buffer zones – rather like adding an extra layer of clothing to trap warm air. The planners were keen to support such an environmentally conscious building.

"Building with our timber frame material of choice – Masonite – was not without its problems," says Laurence. "The first package company pulled out of negotiations because they couldn't find suitable joiners to build an airtight house, and the second one went bust just before we were going to order. Luckily, we found a very good builder who – although lacking in experience with the system – was eager to learn and acted as our project manager. It cost more to erect the frame on site, but we're pleased we took this route because the house is so well built."

The foundations are fairly standard – with one exception: a 20cm layer of Styrofoam insulation material, topped with a 10cm concrete slab, ensures that the floor is extremely well insulated. At this stage, the rainwater tank was also buried in the back garden, and the Masonite I-shaped spruce beams arrived from Sweden. Some recycled wood was used in their production, and the pumped-in cellulose insulation is made from eco-friendly recycled newspapers.

"The softwood windows and conservatory were glazed with low-E glass and, once the house was watertight, work started inside," recalls Monique, who took charge of the finances as well as all of the decorating internally. "By the time we finally moved in, there was still a lot that needed to be finished. It meant carrying our stuff from one room to the next – which kept us very busy!"

Despite another full year of work to complete the house, the couple still thoroughly enjoy living in their new home – concluding that it functions even better than they had expected. In summer, the living room remains relatively cool with the conservatory door closed, and in winter this space is warmed by the wood stove. On sunny winter days and in spring and autumn, the Steijgers can heat the house by opening the conservatory doors and making use of passive solar gain.

"The rainwater tank has not yet been empty and the heat pump supplies us with pressurised water," says Laurence. "After six years in a typical English house with a gravity system, it's so nice to have a powerful shower at the perfect temperature, without having to waste water by running it for ages first."

It was Laurence who built the kitchen, laid floors and made cupboards. He then went on to construct a hen house, using leftover cedar to match the cladding. The couple's quarter-acre site is now a haven for wildlife. It includes a pond and a self-sufficient kitchen garden designed to organic guidelines, with manure sourced from a nearby pigeon racing club, which surely has to be the last word in recycling. ●

**The only element of the original design which is missing is the wind generator. The Steijgers are waiting for a model to come onto the market which is low in noise and vibration. Windows have been kept to a minimum on the north façade to reduce heat loss.**

## FACT FILE

**Names:** Monique and Laurence Steijger
**Professions:** Freelance business analyst and energy consultant
**Area:** Derbyshire
**House type:** Four-bedroom eco house
**House size:** 178m$^2$
**Build route:** Builder and subcontractors
**Finance:** Stage payment mortgage
**Construction:** Timber frames, late roof
**Warranty:** NHBC
**Build time:** June '01-April '03
**Land cost:** £50,000
**Build cost:** £190,000
**Total cost:** £240,000
**Current value:** £300,000
**Cost/m$^2$:** £1,067

**20%**
**COST SAVING**

**Cost breakdown:**

| | |
|---|---|
| Rainwater tank, excavation and fitting | £2,862 |
| Sunslates | £11,226 |
| Cedar cladding | £14,000 |
| Conservatory | £14,400 |
| Masonite, Panelvent, Tyvek membrane | £11,692 |
| Warmcel insulation | £5,368 |
| Chimney and wood stove, incl fitting | £6,839 |
| Heat pump | £1,169 |
| Heat-recovery unit | £750 |
| Architect | £11,419 |
| Building contractor and materials | £110,605 |
| **TOTAL** | **£190,330** |

## FLOORPLAN

**The four-bedroom house includes a home office, an open plan living area, a utility and a cloakroom. Two storage rooms on the north and the double-height south-facing conservatory act as heat buffer zones.**

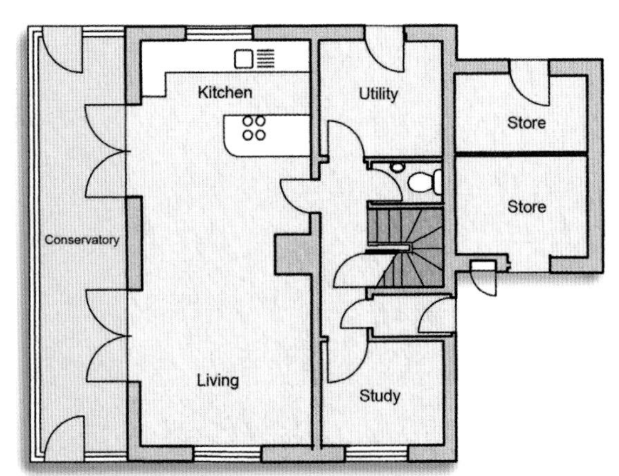


**GROUND FLOOR**



**FIRST FLOOR**

## FEATURES

- **Rainwater recycling**
- **Passive solar gain**
- **Recycled newspaper insulation**
- **Solar panels**
- **Ground-source heat pump**
- **Super insulation**
- **Recycled wood**
- **PV cells**

## USEFUL CONTACTS

**Energy concept and design** Silvercrest Energy & Automation: 01773 861762 www.silvercrestec.com **Architect** Alan Newton Associates: 01773 550008 **Building contractor** Claremont Construction: 01629 823326 **Construction of Masonite, cladding, conservatory, windows and doors** Mitech Joinery: 01773 570577 **Staircase construction and wooden flooring** Nico van der List: (+31) 010 4271351 **Warmcel insulation installation** Pen Y Coed: 01938 500643 **Sunslates** Solar Century: 020 7803 0100 **Accelerator mortgage** BuildStore: 0870 8709991 **HDRA-approved eco garden design** Town & Country Garden Design: 0114 250 0038 **Masonite I-beams** Swelite AB, Sweden: (+46) 0930 14200

# HIDDEN CREDENTIALS

**It may look run of the mill at first glance, but the Victorian house of BBC Two's *No Waste Like Home* presenter Penney Poyzer and husband Gil Schalom is in fact a standard bearer for how to transform old houses into cutting-edge eco-friendly homes.**

WORDS: CLIVE FEWINS PHOTOGRAPHY: JEREMY PHILLIPS

The home of TV presenter and environmentalist Penney Poyzer and her architect husband Gil Schalom looks like all the others in the row of three storey Victorian semis, a mile or so from the centre of Nottingham. Aside from the array of solar collectors on the roof, there is little evidence to suggest that this is the self-styled Nottingham Eco House.

Look a little closer, however, and you will observe that the copper drainpipe carrying rainwater from the roof has a filter at the base. The pipe connects to two huge tanks in the cellar that store up to 2,500 litres of filtered rainwater, which is used in the low-flush WCs throughout the house.

Step inside the front door and you immediately experience a fresh atmosphere. An earth plaster covers most of the downstairs walls, including those covered with a heavy layer of zero-ozone-depleting foam insulation board.

"The house had been student lodgings and had had a very cheap makeover two years before we bought it in 1998," says Gil. "It was pretty desperate to the point of internal toxicity. It had been coated throughout with a polymer-based emulsion paint and many of the ceilings were covered with polystyrene tiles."

"It was totally inappropriate for a house built as a breathing structure with traditional nine-inch brick walls," says Penney, who presented last summer's eight-part BBC Two series *No Waste Like Home*.

Penney and Gil's house demonstrates many of the things she was trying to promote in the series, especially the reduction of waste – particularly that of precious energy.

Apart from the solar collectors, there are rooflights with low-emissivity glass ➤

**Aside from the array of solar collectors on the roof, there is little evidence to suggest that this is the self-styled Nottingham Eco House.**

we are
but
mutants

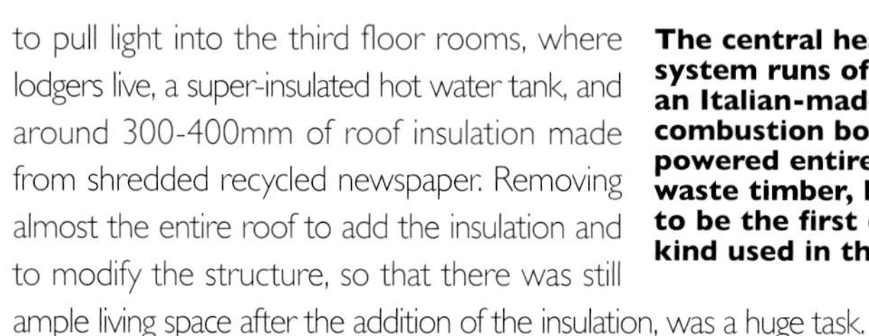

to pull light into the third floor rooms, where lodgers live, a super-insulated hot water tank, and around 300-400mm of roof insulation made from shredded recycled newspaper. Removing almost the entire roof to add the insulation and to modify the structure, so that there was still ample living space after the addition of the insulation, was a huge task.

**The central heating system runs off an Italian-made combustion boiler powered entirely by waste timber, believed to be the first of its kind used in the UK.**

Inside the front façade of the house, there is 100mm of ozone-friendly insulated dry-lining – this has also been installed at the corners, where there was a high level of heat leakage. On the side and north-facing rear wall, where it is less important to show the period brickwork, 150mm exterior wall insulation has been applied. As the walls are solid brick there was no possibility of cavity insulation, so rigid insulation boards with an insulated render finish were applied, bought at a reduced rate of five per cent VAT on the £12,000 cost of the job.

Double and triple timber-glazed timber units have replaced many of the PVCu windows. "PVCu windows are bad news from an environmental point of view because of the problems with disposal," says Penney. Where they are not replaced it is because some were newly installed before they bought the house, and also for financial reasons.

In addition to all these features, Penny and Gil have installed a completely new central heating system that runs off waste timber, meaning that fuel costs them nothing. "There are so many construction sites round here that I do not think we'll ever run short of free fuel," says Penney. "A number of builders drop off their scrap pallets, and we are also great skip raiders!"

Combustion takes place in a manual boiler. It is an Italian product, and Penney and Gil believe it to be the first domestic application in the UK. "It is specifically designed for biomass, and for heating water," says Gil. It works as a top-up to the solar water heating system.

One of the most fascinating eco features is the brick and concrete

composting chamber in the cellar, which has a raft inside composed of coir (fibre from coconut husks) and a large colony of worms. The contents only need to be removed about once every three years and can be used as garden compost.

"The brick frontage is unchanged. The whole idea has been to test new products to demonstrate that other people on relatively low incomes – perhaps in smaller terraced houses – can do this and make significant energy savings," says Penney.

"So far the house has cost about £85,000 to renovate and we still have to replace the kitchen. We reckon that if we had not used so many green techniques we could have done it for about half this figure," says Gil. "But our motivation is more for environmental reasons than to obtain rapid payback – we do not expect to reap the full economic benefits for 10 years or so. However, in terms of water and energy savings we are now paying

The curved shower enclosure is made from bendable Finnish birch ply. It was inspired by famous architect Gaudi and designed by Penney and Gil.

less than £500 a year in electricity for top-up water heating when the solar panels do not produce enough hot water and for all other purposes. Not too bad when you consider that this is a $240m^2$ house.

"There is also the equity point. We bought a very run-down property in 1998 for £84,000 when similar properties near here were fetching £120,000. It is now worth over £300,000 so we have not exactly cut off our noses by using green methods. We think we have also shown that, despite the undoubted need for thousands of new homes in the UK, there is huge potential for updating the existing housing stock – 60 per cent of which dates from before 1919."

As a result of all of Gil and Penney's efforts, the house has been recognised by the Government-sponsored Energy Saving Trust as one of the most advanced low-energy retrofits in a house in the country and is being used as a model in two of the Trust's Best Practice manuals.

"None of our house's features in its own right is revolutionary, but together they amount to a huge energy saving," states Gil. "Because of this we are close to being a zero-$CO^2$ house."

## FACT FILE

**Names:** Gil Schalom and Penney Poyzer
**Professions:** Architect and writer and green campaigner
**Area:** Nottingham
**House type:** Three storey, four bedroom late-Victorian semi
**House size:** $240m^2$ incl. cellar
**Build route:** Selves as main contractors
**Construction:** 9" brick
**Finance:** Private plus £114,000 Virgin mortgage
**Build time:** Six years
**Land cost:** £85,500
**Build cost:** £169,000
**Total cost:** £254,500
**House value:** £325,000
**Cost/m²:** £354

## 22%
**COST SAVING**

**Cost Breakdown:**

| | |
|---|---|
| Fees and preliminaries | £1,000 |
| Solar collectors | £3,500 |
| NSM Building | £10,000 |
| Jason Lewis, builder | £20,000 |
| Heating and plumbing | £17,000 |
| External render | £12,000 |
| Rainwater collection system | £3,000 |
| Cellar, alterations & compost system | £5,000 |
| Roof, cellar and other insulation | £4,000 |
| Sanitaryware | £1,000 |
| Electrics | £3,000 |
| Other materials | £2,500 |
| Miscellaneous | £3,500 |
| **TOTAL** | **£85,500** |

## GREEN FEATURES

- **Biomass heating system**
- **Composting chamber**
- **Rainwater harvesting**
- **Additional insulation**
- **Solar panels**
- **Natural paints**
- **Low-emissivity glass**

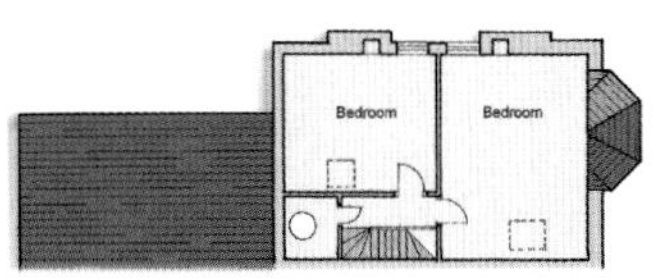


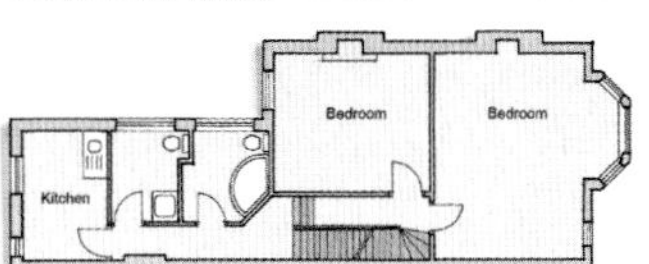


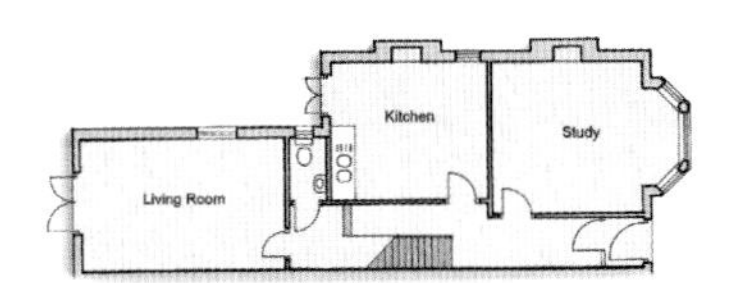


## FLOORPLAN

**The terraced house has a conventional layout, with extra self-contained accommodation for the lodger on the first floor.**

## USEFUL CONTACTS

**Breather membranes** MAD Proctor: 01250 872261 **French windows and oak windows** Environmental Construction Products: 01484 854898 **Rainwater harvesting system** The Green Shop: 01452 770629 **Eco paints** Holzweg & Aquamarijn, Construction Resources: 020 7450 2211 **Other natural paints:** Environmental Construction Projects: 01484 854898 **Zero formaldehyde MDF** Medite: 01702 619044 **Slate** JA Stephens: 0115 941 2861 **Low-emissivity glass** Interplane Safeheat: 01457 837779 **Floor insulation** Jablite: 020 8320 9100 **Insulation** Warmcell: 01685 845200; Second Nature: 01768 486285 **Roof insulation** Rockwool: 01656 862621; Regupol: 020 7450 2211; Bitvent rigid breather board: 01685 845200 **Breathing roof felt** Proctor: 01250 872261 **Boiler** Mescoli: 01825 890140 **WCs** Elemental Solutions: 01981 540728; Ideal-Standard: 01482 445886 **Heat recovery fans** Vent-Axia: 01293 526062 **Composting chamber** Elemental Solutions: 01981 540728

# NATURAL BEAUTY

Marion and Francis Chalmers have built a new home in Norfolk that exemplifies the principles of sustainable development.

WORDS: HAZEL DOLAN
PHOTOGRAPHY: PHILIP BIER

This house feels right," says Marion Chalmers of her rammed earth and timber-framed Norfolk home. "The light is absolutely fantastic. It's semi open plan, but we have glass to separate spaces. It is unique and it works beautifully."

For Marion and her husband, Francis, the idea of a self-build using natural materials had long appealed. Ten years of visiting self-build exhibitions and two actively searching for a plot ended in July 1998, when they came across a prefab on a quarter of an acre site at the edge of a pretty coastal village.

They loved the site, the village and the North Norfolk environment. The fact that there was an existing habitable house on the plot was an added bonus. It meant they could live on the site while they finalised their plans.

It also had one critical feature very important to them: it had an east entrance and faced almost due east. This was key to their plan to use the ancient Indian system of Vedic architecture, inspired by Maharishi Sthapatya Veda, which requires houses to have their main entrance facing due east. Described as 'building in accord with natural law', it deals mainly with a house's orientation, the placement of rooms, and harmonious proportions of the house and elements within it. In general, rooms are positioned to get sunlight at the time they are used most.

They were up against at least eight serious bidders at the auction. "We'd worked out our final price and weren't going to go above it," says Marion. "I was bidding and trying to look very nonchalant, as if I could go on forever, when actually I knew that we were very near our limit. Someone else then put in a bid to our limit and I didn't look at Francis. I just added another thousand, and it stopped there, thank goodness! I don't know what we would have done if it had gone on.

"We bought it without knowing what we were going to put on the site. We knew from the planners that they were happy for it to be replaced, but we had no idea what we were going to be able to get."

Friends recommended Mike Brackenbury, a former North Norfolk District Council conservation architect, and there was an immediate rapport. "He was enthusiastic about Sthapatya Veda and he loved the spot. We could talk to him and felt he would give us what we ➤

**The traditional style kitchen is from John Lewis of Hungerford (0700 278 4726).**

"THE LIGHT COMES IN FROM THE TOP, RIGHT THE WAY THROUGH THE OPENNESS AND THE HOLE IN THE CEILING AND THAT VISUALLY TIES IT TOGETHER"

wanted. Light and spaciousness were the main priorities. I remember saying to him: 'I don't like open plan but, actually, I think I want it here!' It was what the site dictated."

His view was that the space required a structure with a simple, solid presence. He also suggested earth building as a form, and recommended conservation builder Tim Hewitt to do it.

Over nine months they refined the design, but the philosophy stayed the same: to use natural materials, sourced locally, and local craftsmen, tradespeople, suppliers and labour, wherever possible.

From their first meeting with Tim Hewitt, they felt confident. "One of the main things about Sthapatya Veda is precision, so you have to have very precise dimensions. Things have to be very straight and orientated properly, and we felt he was very enthusiastic. He had done a lot of clay lump building, and rather like Mike, he felt he loved what he was doing, but saw no reason why they should keep on producing traditional houses."

Planning permission was granted in June 2000, but it took a further ten months to get their proposal through building regulations. The lack of data on the behaviour and properties of rammed earth proved a drawback, but the building control officer at North Norfolk District Council was encouraging and supportive.

"With building regulations they have very strict ways of measuring things which are really only suited to certain materials, and earth walls just don't come into that category, so they don't do well in terms of their measurement of thermal capacity. Of course, in practice they function absolutely brilliantly, because they absorb the warmth from the sun and then they gradually give it out. When it's cooled down outside, they are continuing to give out the warmth inside. But to get through building regulations, we had to either have very small windows, or do what we did, which is put a timber frame upstairs and pack it full of insulation, which passes and exceeds the current Building Regulations and averages out over the house."

Another deciding factor in choosing a timber frame for the second storey was time. Earth building is a lengthy, labour intensive process, and the Chalmers felt it would not prove practical for a house on this scale. "We've ended up with a house which is different upstairs from downstairs in terms of construction and finishes, but I think the way it has been designed and finished makes the whole thing tie together beautifully. The light comes in from the top, right the way through the openness and the hole in the ceiling and that visually ties it together."

Green materials were important both as a personal choice and as a recommendation of Sthapatya Veda. Rather than polyurethane, they used ➤

**The house is designed around the principles of Vedic architecture, which concentrates on the qualities and orientation of light and space. A lightwell pierces the main living space from a roof lantern through to the main reception area.**

**Acoustic insulation between floor and ceiling, a wood shaving product called Silencio, provides the necessary sound insulation for a house without carpets.**

**Long, unsupported roof spans required complex joist detail, with one large steel beam and two smaller Douglas fir beams to cater for the loadings involved.**

blown clay granules for insulation beneath the floor slab, Warmcell natural cellulose was packed between the first floor rooms, and they used plant based paints and stains for the interior and linseed oil based paint for the exterior.

The prefab was demolished and although they began by living in a caravan on site, the Chalmers and their sons, James and Michael, soon realised it was impractical, and decamped to a nearby rented house. Work began on the extra wide foundations needed to support the 2.7m high earth walls, which were built in sections.

"You put up wooden shuttering in the form of the house and mix up the mixture in the concrete mixer – we used hoggin, sand, lime, a little bit of white cement to help with the strength and get through building regulations, and some chalk for consistency," explains Marion. "It is poured into the shuttering, then rammed down with a compressor-driven sand rammer. It all happens quite quickly. You have one person mixing, wheeling over, pouring in, someone ramming down and then the shuttering is taken down and you are left with earth walling standing.

"Tim Hewitt is a craftsman and worked for a long time to get the finish and colour right. This would be the visible finish, not just the material. It's very lightly glazed with a slight wax and seal, which gives it that very light sheen. When it catches the light, it goes a beautiful gold colour."

The timber frame is clad with natural materials: a zero-formaldehyde, breathable pressed fibreboard, Sarket, then a gap, battens and red cedar boarding. Cedar was chosen both for its attractiveness and natural preservatives, which give it an untreated life of over 60 years.

"One of the most exciting things was when the roof was put on," says Marion. "It's quite an unusual structure because of the rooflight in the middle. You could see the crane, with this structure to support the huge lantern on it, swinging out over the field, and then being positioned so carefully."

Solar roof panels are designed to heat 80 per cent of their hot water. In winter a highly efficient condensing boiler takes over. "One of the ways this

**Tall glazed doors and wide windows frame the view of the garden and fields beyond.**

house works so well is that it benefits enormously from solar gain, so that even in winter, if the sun is shining, the heating doesn't need to come on," says Marion. "This means in the evening the underfloor heating hasn't been on and that is where the woodburning stove comes in. It's lovely to have a genuine reason to put it on."

For the Chalmers family, designing and building their home has been a very personal and rewarding experience. "We did, of course, spend more than we wanted to spend," says Marion. "I think our perspective on what we were building changed quite quickly. Once we realised we were going to get the sort of house we wanted, and that we would probably live in it ad infinitum, our view altered.

"In most houses you heat the air inside and you don't think about what is beyond the space within the walls. I think when you have got natural materials and you have thought about the light and the movement in the house, you feel quite differently. You feel you are living in the whole house and simultaneously enjoying the surroundings to the full." ●

## FACT FILE

**Names:** Marion and Francis Chalmers
**Professions:** Transcendental meditation teacher and maths teacher
**Area:** Norfolk
**House type:** Detached four-bedroom house
**House size:** 200m²
**Build route:** Builder and local subcontractors
**Construction:** Rammed earth walls, timber-framed first floor, clay pantile roof
**Finance:** Private
**Build time:** July '01-end of '03

**Land cost:** £106,000
**Build cost:** £270,000 (after VAT reclaim of approx. £20,000)
**Total cost:** £376,000
**House value:** £650,000
**cost/m²: £135**

**42%**
**COST SAVING**

## FLOORPLAN

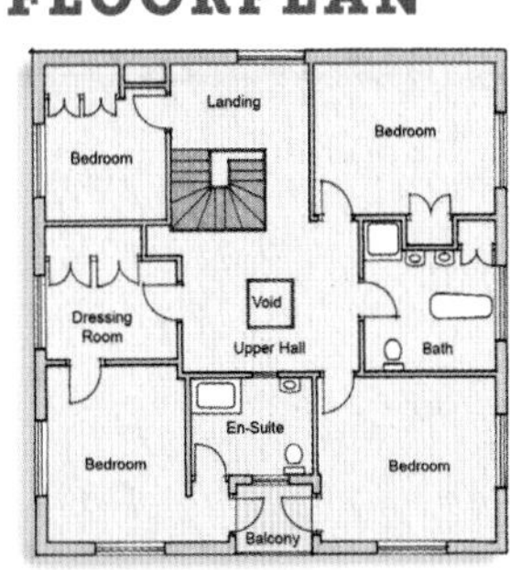

**FIRST FLOOR**

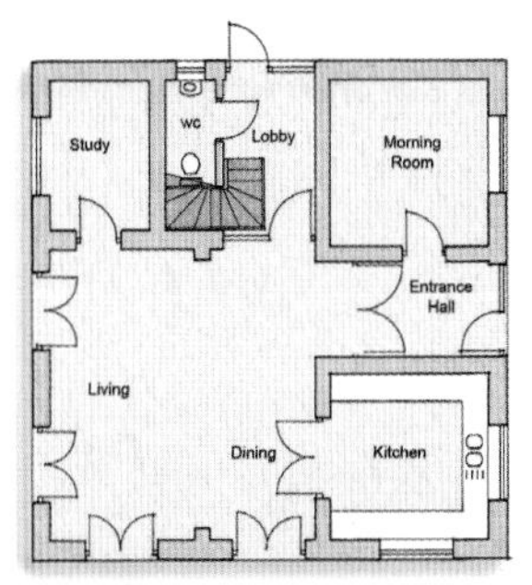


**GROUND FLOOR**

**A triple height void pierces the main living space, around which the rooms run, therefore cutting down on corridor space.**

## USEFUL CONTACTS

**Design architect** Michael Brackenbury: 01493 667888; **Technical architect** John Bradley at Apt Architecture: 01263 515009; **Structural engineer** Alan Gentry: 01263 834949; **Maharishi Sthapatya Veda consultant** Bob Glover: 07950 030642; **Main builder and earth building specialist** Tim Hewitt, conservation builder: 01692 650162; **Carpenter** Stuart Paramor: 01263 833034; **Windows and doors** North Norfolk Joinery: 01263 515696; **Staircase** Cullum & Clarke: 01603 860564; **Rooflantern** R. Wadlow: 01603 308928; **Underfloor heating, solar panels and boiler** Eco Hometec: 01302 722266; **Suppliers of Silencio underfloor acoustic insulation and Sarket** Hunton Fibre: 01933 682683; **Specialist plasters, external insulating render** Telling Lime Products: 01902 789777; **Paint** Ecos Paints: 01524 852371; **Linseed paints** Holkham Paints: 01328 711229; **Cedar** Capricorn Timber: 01283 821110; **Chestnut flooring** Petersons Natural Floorings: 01263 761329; **Kitchen** John Lewis of Hungerford: 0700 278 4726.

The vaulted ceiling with exposed rafters, covered with tongue and groove boarding is a feature of many of Fleming Homes' designs.

Aileen and Dave Downie have built an upside-down-style new home on a tight urban site that costs just £37 a year to heat.

# HIGH STYLE, LOW ENERGY

WORDS: HEATHER DIXON PHOTOGRAPHY: DAVE BURTON

In a road of traditional 1930s semis, Aileen and Dave Downie's unconventional self-built house – with its high elevations and cedar wood cladding – could have stood out like a sore thumb. But by pushing the property back against two high brick boundary walls and leaving a beautiful old beech tree intact at the front, they managed to persuade York planning department that the house would be barely visible from the road.

Imaginative window placement means their closest neighbours are not overlooked, and a private terrace over the garage offers extra garden space in a plot where outdoor space is at a premium.

But where the Downies have become the real envy of their neighbours is over the heating bills. It costs Aileen and Dave just £37 a year to keep their ecologically-friendly home at a steady 70°F, thanks to very high levels of insulation and a controlled whole house ventilation system with heat recovery.

"We are both into energy conservation, renewable materials and the health benefits of eco-friendly living," says

Easy access to all areas was key to the internal design, and the kitchen was deliberately positioned between the dining and living spaces for this reason.

Aileen. "We knew we wanted a timber-framed house, but we were keen to experiment and push the boundaries. We put a pack together outlining our ideas for the design, use of space, and so on and sent it out to 10 companies. Only three showed an interest."

They settled for Fleming Homes, who not only accommodated the Downies' design, but also encouraged them to consider new ideas which would suit their individual needs. "We liked the fact that they were so flexible," says Dave.

With limited help from a 'forward-thinking' architect, the Downies submitted their plans, but they were initially rejected on the grounds that there were too many windows overlooking neighbouring properties. A modified plan, with narrow, floor level windows replacing those which directly overlooked other houses, was subsequently accepted with encouraging feedback.

"We were told that we had obviously spent a long time planning the house and that it was a good design, which was wonderful when you hear of so many negative stories about planning departments," says Dave.

There was just one potential setback. Although outline planning permission had existed on the plot since the 1930s, the area had since become recognised as a site of archaeological interest. In order to proceed, therefore, the Downies had to carry out a desktop study of the history of the land going back to the early 1900s, and to agree to an archaeologist supervising the digging of the foundations. To create as little disturbance to the site as possible, Aileen and Dave ordered a structural survey and opted for raft foundations, rather than digging deep into the ground. The cast concrete raft, reinforced by steel mesh, incorporates both the foundations and oversite slab in a single entity which spreads the weight of the building evenly across the ground beneath.

"BY HAVING THE SITTING ROOM, KITCHEN AND DINING AREA ON THE FIRST FLOOR, WE ARE ABLE TO MAKE THE MOST OF THE HIGH, OPEN ROOF SPACE"

The timber frame kit was erected to a weathertight shell in just eight weeks and the first-floor cedar cladding – which is treated with Danish oil to prevent it turning 'silver' – was bolted into place over a triple layer of Rockwool insulation. On the ground floor, the frame was ➤

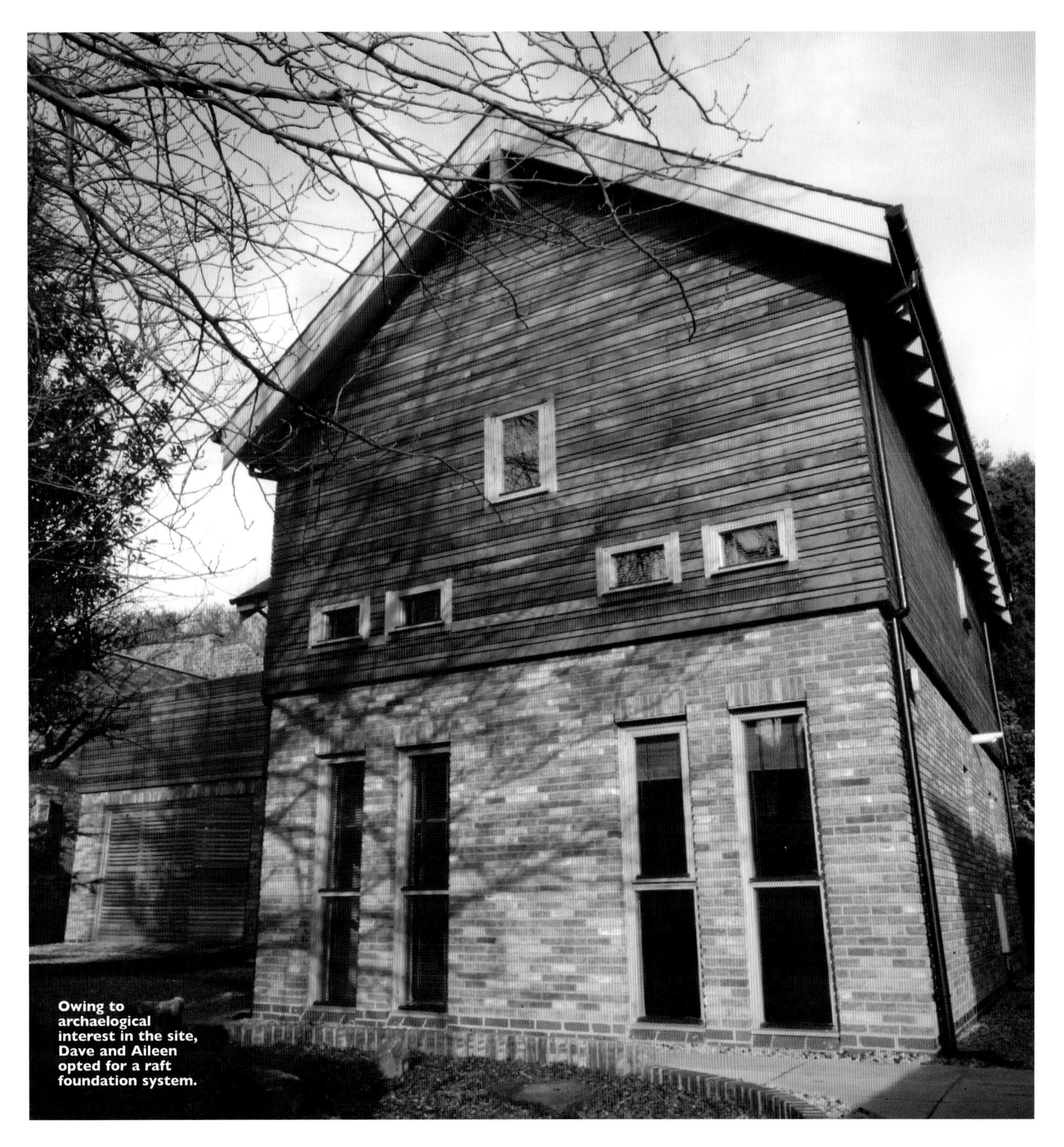

Owing to archaelogical interest in the site, Dave and Aileen opted for a raft foundation system.

"THE TIMBER FRAME KIT WAS ERECTED TO A WEATHERTIGHT SHELL IN JUST EIGHT WEEKS"

clad in Friston Multi brick to tie in with neighbouring housing.

Internally, Dave had originally wanted to have a conventional ceiling with a room in the roof above. However, Fleming prefer to leave the roof space open to create a vaulted ceiling, with exposed rafters, covered with double tongue and groove boards and insulated on the outside. Dave liked the idea and so they took the company's advice and have left the roof space open – with the option of installing cross beams to form a room in the roof at a later stage, should they ever need the space.

As the build progressed, Dave became so involved in managing the project and taking on as much of the labour as he could, that he gave up work to invest all his time and energy into the house. This included 80% of the electrics, painting, some joinery work – including laying the Canadian oak floor – plastering and helping to secure the pipework for the underfloor heating in the first floor living area, which had to be clipped into place from underneath. "It took me and the plumber a whole week to install the piping for that one area," says Dave. "If I did it again I would create a chipboard deck and fix the pipework from above, then build floors and ceiling around it that way."

Both Aileen and Dave were determined to make the most of natural light as part of their aim to design out 'Sick Building Syndrome'. "The UK weather is often very dull so we wanted to maximise on natural light without losing heat through too much glass," says Aileen. "It was one of the reasons we decided to design the house with the main living areas on the first floor and the bedrooms on the ground floor, where they would be cooler in summer and cosier in winter.

"By having the sitting room, kitchen and dining area on the first floor, we are able to make the most of the high, open roof space, draw as much natural light as possible into rooms where it's most needed and develop a more efficient heat flow system."

The Downies installed a weather-compensating Vaillant boiler in the smallest utility area possible, over a washing machine and conveniently next door to a shower/sauna room. The highly insulated and airtight house is ventilated by means of a mechanical system rather than via trickle vents or airbricks. The system has a heat exchanger which recovers up to 75% of the ➤

The open tread steel-and-wood stairs have exposed bolts. The wire for the balustrading was imported from Switzerland.

**Together with large timber windows made in Sweden, the series of rooflights helps to ensure that the first-floor living space enjoys plenty of natural light.**

energy from the warm, damp, stale air extracted from the kitchen and bathrooms, to preheat the cold incoming fresh air.

Lighting was another major consideration. "We believe that a house should be designed so that the building itself creates a visual statement, rather than rely on the furniture that goes in it," says Aileen. "We wanted the lighting to work with that. We didn't want table lamps everywhere but we liked the idea of subdued and disciplined lighting to highlight selected areas of the house."

Easy access to all areas was key to the internal design, and the kitchen was deliberately positioned between the dining and living spaces for this reason. The open tread steel and timber stairs with exposed bolts are Dave's pride and joy.

"I wanted to incorporate a serious piece of steelwork into the design," he says. "I draw inspiration from industrial buildings, so I liked the idea of bringing heavy duty materials into the inside of the house. We approached a firm in Switzerland to provide the wire for the balustrading because we couldn't find what we wanted in the UK. We sent them dimensions and the wires were delivered the next day. They were incredibly efficient."

The house is now more or less complete, but Aileen and Dave have become so hooked on the idea of building ecologically friendly houses that they have decided to turn it into a specialist business – in Canada.

"We have been so encouraged by people's reactions that we want to design and build more bespoke contemporary houses like this," explains Aileen. "It doesn't just make ecological sense, it's terrific fun as well!"

## FACT FILE

**Names:** Aileen and Dave Downie
**Professions:** Manager training and service engineer
**Area:** York
**House type:** Three-bedroomed detached
**House size:** 186m²
**Build route:** Self-managed with sub-contractors
**Construction:** Timber frame with part brick, part cedar cladding and slate roof
**Warranty:** Premier Guarantee
**Sap rating:** 95
**Finance:** Self financed
build time: April '02-June '03
**Land cost:** £60,000
**Build cost:** £190,000
**Total cost:** £250,000
**House value:** £395,000
**Cost/m²:** £1,021

**37%**
**COST SAVING**

**Cost breakdown:**

| | |
|---|---|
| Planning and fees | £10,000 |
| Build costs, inc. foundations and boundary walls | £65,000 |
| Timber frame and erection | £55,000 |
| Service connections and installations | £12,500 |
| Timber and tile flooring | £11,000 |
| Kitchen and bathrooms | £12,000 |
| Landscaping/clearance | £6,500 |
| Heating, heat recovery, plumbing and electrics | £18,000 |
| **TOTAL** | **£190,000** |

## FLOORPLAN

**The main living spaces are situated on the first floor.**

## USEFUL CONTACTS

**Architect** John Baily: 01845 578568; **Timber frame manufacturers** Fleming Homes: 01361 883785; **Timber Flooring** Jorwood Flooring: 01904 792759; **Brickwork** Kevin Pallister: 01904 634219 and Phil Wilbraham: 01904 658869; **Steelwork** PCB Fabrications: 01904 607073; **Brick supplier** Blockmere Brick & Tile Company: 01423 324626; **Underfloor heating** Nu Heat: 01404 549770; **Blinds** York Blind Company: 01904 416389.

**The Herberts' eco bungalow is timber-framed, made from renewable softwoods, clad in American oak and Cotswold stone to fit in with the Conservation Area.**
**Left: Sliding glass doors are shaded by overhanging eaves, which help to reduce glare from the summer sun.**

# GREEN HOUSE IN THE GARDEN

Pat and Dennis Herbert have built themselves an environmentally friendly retirement bungalow in the garden of their former home.

WORDS: DEBBIE JEFFERY PHOTOGRAPHY: TONY WARD

"Everyone's talking about the environment and saving energy, and we decided that our new house should be as green as possible," says Dennis Herbert of his timber-framed Gloucestershire home. "We had lived in our 1950s house for 22 years and our children grew up there, but when they left home it was too large for just the two of us. We didn't want to move out of the village, so we decided to build in the garden to the south of the existing house."

The Herberts live in a Conservation Area and realised that they would need a sympathetic architect to design their new home. "We read about the 'Architect in the House' scheme, which was how we found Bruce Buchanan," says Pat. Every year during Architecture Week the Royal Institute of British Architects promotes this scheme to encourage members of the public to register with RIBA to be matched with an architect in their area. The architect then visits them at home to give a consultation in exchange for a minimum donation of £25 to Shelter, the UK's leading housing and homelessness charity.

Specialising in ecological, sustainable, energy-efficient designs

"AT FIRST WE THOUGHT ABOUT BUILDING ON TWO STOREYS, BUT DECIDED THAT A SINGLE-STOREY BUNGALOW WOULD BE MORE PRACTICAL FOR OUR RETIREMENT"

for new houses, the Buchanan Partnership in Cheltenham proved the ideal practice to help the Herberts achieve their goal. "When I first met Dennis and Pat they were very keen to find out more about building a green house," says architect Bruce Buchanan. "I gave them a reading list and suggested that they visit the Centre for Alternative Technology in Wales for inspiration on living more sustainably — and then didn't hear from them again for several months. To their credit they did a huge amount of homework in that time so that, when I next saw them, they had firm ideas about what they wanted to include in the house."

Bruce developed ideas from a selection of designs, and energy-saving measures include southerly orientation to maximise passive solar gain, high levels of recycled newspaper insulation, triple glazing and low-E glass, a thermal heat-exchange pump for space heating, a solar panel for hot water heating and PV cells for electricity generation. Every detail has been considered – from energy-efficient light fittings to eco-friendly paints – and there is a Passivent natural ventilation system which provides automatic humidity control.

Planning permission was granted for the bungalow in June 2003, and building work began six months later. "At first we thought about building on two storeys, but decided that a single-storey bungalow with no stairs would be more practical for our retirement, as well as making planning easier because it avoided overlooking our neighbours," says Dennis. "Originally, a basement garage and utility room had been planned and approved by the Council but, when the builders started work on site, they discovered a seam of running sand which made it impractical. It was disappointing, particularly as we had hoped to recycle rainwater from the roof and collect it in an underground tank."

Access onto garden plots can sometimes prove difficult but, with a right of way over a lane leading into their garden, the Herberts had no such problems, despite a minor dispute with the neighbours. They chose to build using a timber frame supplied by Taylor Lane of Hereford, and insulated with Warmcel recycled newspaper insulation. The structure was clad in local stone and untreated American oak, which will weather to a silver grey over time.

The Herberts lived in their old house throughout the build, and named their new home 'Greengages' to reflect the fact that it is an eco house. "We ➤

"WE HAVE GENERATED 3,600 UNITS OF ELECTRICITY AND HAVE EXPORTED 2,085 OF THESE BACK INTO THE NATIONAL GRID, FOR WHICH WE ARE PAID"

are saving money on our power bills because of the photovoltaic panels on the roof," says Dennis. "Since we have lived in the house these have generated 3,600 units and have exported 2,085 of these back into the national grid, for which we are paid. Additionally, we were given a 50% grant from the DTI towards the initial costs. The ground source heat pump runs the underfloor heating, and we have around 300 metres of water piping installed under the garden – about two metres down – which draws heat from the ground. There was a £1,200 grant for this and we also had a £500 grant for the solar panels, which helped to compensate for the outlay."

The Herberts' new bungalow may be designed as a low-maintenance home for retirement, but it incorporates luxury features such as a steam shower, multi-room hi-fi and mood lighting. Maple kitchen units are fitted with hardwearing Corian worktops, and roller shutter doors conceal appliances to either side of the stainless steel range cooker.

**Top left: Exposed glulam beams add character to the spacious living room. Top right: Roller shutter doors conceal appliances.**

"When Bruce was designing the house we told him that we wanted high ceilings which would enhance the impression of space," Pat explains. "Most of the south-facing wall is glazed, and we have full height sliding glass doors in the lounge which came from Sweden, but there are no windows to the north at all. To compensate for the lack of windows to the rear of the plan, three sunpipes drop daylight down into the bathrooms and hallway from the roof through reflective tubes. There are also rooflights in the kitchen and dining room, and we commissioned a stained glass window for the lounge which casts coloured light through into the kitchen. All of this means that we have very bright rooms but don't feel at all overlooked."

Despite concerns that leaving their old house would be a wrench after living there for so many years, the Herberts settled immediately into their new home, and enjoy the fact that it is so inexpensive to run. The log burner in the sitting room was installed as a feature, but has yet to be lit thanks to the energy-efficient nature of the building. "People are surprised when we tell them about the various features because so many of them are hidden from view. We did go over budget, and paying for items such as the ground source heat pump certainly bumped up the final bill," says Dennis, "but the payback starts immediately and we like to think that the house will be enjoyed by future generations, who will also be helping to reduce energy consumption."

## FACT FILE

**Names:** Pat and Dennis Herbert
**Professions:** Retired
**Area:** Gloucestershire
**House type:** Two-bedroom eco bungalow
**House size:** 175m$^2$
**Build route:** Main contractor
**Construction:** Timber frame, oak and stone cladding, clay tiles
**Warranty:** Architect's Certificate
**Finance:** Private
**Build time:** Dec '03-Nov '04
**Land cost:** Already owned (valued at £100,000)
**Build cost:** £370,000
**Total cost:** £470,000
**House value:** £500,000+
**Cost/m²:** £2,114

**6%**
**COST SAVING**

"PEOPLE ARE SURPRISED WHEN WE TELL THEM ABOUT THE VARIOUS FEATURES BECAUSE SO MANY OF THEM ARE HIDDEN FROM VIEW"

## FLOORPLAN

**The lounge/dining room leads into a separate kitchen, and there is a study, WC, utility room, dressing room, two bedrooms and two bathrooms on the same level. All habitable rooms face south.**

## GROUND SOURCE HEAT PUMPS

Heat pumps can be an ecological way of providing domestic hot water and central heating. A refrigerant is pumped through the ground in pipe coils where it is warmed by the latent solar energy. The heat pump then cools the refrigerant and the energy extracted is used to heat a thermal store. Powered by electricity, a heat pump can produce between 3-4kW of usable heat for every 1kW of electricity consumed. They are most ecological – and economical – when powered by green electricity from solar or hydro. Typically, a system will cost between £4,000 and £6,000, with a household grant of £1,200 offered regardless of the system size. The expected lifespan is at least 30 years.

## USEFUL CONTACTS

**Architect** Bruce Buchanan, Buchanan Partnership: 01242 541200 **Timber frame** Taylor Lane: 01432 271912 **Triple glazed windows and beech doors** Swedish Window Co. Ltd: 01787 467297 **Ground source heat pipes and heat pump** Geothermal Heating (International) Ltd: 024 7667 3131 **BP Solar photovoltaics (generating electricity)** Beco Batteries Ltd: 01803 833636 **Underfloor heating** Mas Heating Systems Ltd: 01926 814969 **Eco paint and solar panels (20 Solamaz direct flow tubes for hot water)** The Green Shop: 01452 770104 **Electrical contractor** SE Barnard: 01242 680207 **Home cinema and music system, data, and lighting system** The Multi-Room Company: 01242 539100 **Sunpipes** Monodraught Ltd: 01494 897700 **Warmcel 500 recycled newspaper insulation** Excel Industries: 01495 350655 **Stained glass window** Paul Phillips: 01527 878167 **Floor tiles** Topps Tiles: 0800 783 6262 **Redland roof tiles** Lafarge Roofing: 01306 872000 **Stone** Farmington Natural Stone Ltd: 01451 860280.

**The house design incorporates passive solar principles, high insulation standards and as many natural materials as possible. There is a rainwater recovery system and drainage is separated for greywater recovery – the WCs are plumbed to use recycled water.**

# A GREEN HILL FAR AWAY

## David and Ellie Austin have built a new eco-friendly home that fits into its sloping hillside setting very well, thanks to the use of natural stone and slate.

WORDS: MARK BRINKLEY PHOTOGRAPHY: MARK WELSH

Floris House is the handiwork of architect David Austin and his wife Ellie. It sits rather wonderfully on a steep hillside in the Cotswold town of Nailsworth in Gloucestershire; it manages to blend in subtly with the surrounding limestone buildings and yet also stands out as something rather different. Its curved central space and planted roof stand as testament to the fact that this is no ordinary self-build. The Austins have a green approach to living and the house reflects this in every way.

Their working lives revolve around the unique community centre of Ruskin Mill. It acts as both a craft centre and workshop area for the town but also, more importantly, as an educational centre for young adults with learning difficulties. Ellie works there full time, developing the landscape and supervising the students in the gardens, while David runs his practice from the Mill. "Originally Ellie got a job here in 1991," says David "and our involvement just grew from there. We loved living here, walking to work and to the shops. The stone houses are beautiful, although there are undoubtedly drawbacks to living in the old ones."

The plot on which Floris House sits has not been easy to develop. It is cut

into a steep hillside and its road frontage is along the top edge. Although easily wide enough to take a conventional house, the depth of the plot is just 15 metres on a one-in-three slope. The house is built into the slope – requiring 150 tonnes of soil to be excavated and carted away to get something flat enough to build on. The north road frontage only has three windows and a front door. "About 20 years earlier two concrete bases had been built then abandoned because of the difficult access. They were too big to remove and so our design had to incorporate them," says David.

"We built a model to more easily explain the design but still had terrible trouble getting detailed planning permission," explains David. "There was permission for a very conventional four-bedroom house with access from down below. We thought our scheme was a clear improvement using

"WE THOUGHT OUR SCHEME WAS A CLEAR IMPROVEMENT, BUT THE PLANNERS WERE DETERMINED TO REFUSE IT"

**Floris House has an extraordinary living room with a 24-spoke ceiling and a fireplace that was made for them locally from pressed steel.**

The wonderful curved staircase is made out of concrete lintels built into the blockwork walls.

natural materials but the planners were determined to refuse it. It's a brownfield site, well inside the development envelope, it follows the building line, there was only one objection (which was easily accommodated – our scheme actually impacted on them far less than the original one they had accepted) and the town council approved – but still the planners were of a mind to turn us down! Our scheme ended up in front of Stroud District Council where it lost by just seven votes to six. Not disheartened, we immediately appealed against this decision. The appeal was heard four months later and we succeeded in overturning the refusal – so we were able to build to our original design."

Planning was not the only red tape that caused problems. "The VAT reclaim was the most harrowing thing in the whole self-build process," says David. "They are so strict. You can't even buy tools and invoices that have tools included have to be broken down. We ran out of money and had to do the VAT reclaim early in order to free up £10,000. I have no idea why you can't do quarterly reclaims like normal businesses. In fact there are many items which we have yet to purchase which we could in theory reclaim the VAT on but we won't be able to because we have already made our one and only claim."

David undertook a great deal of the building work himself. "From May to September I took time off work to concentrate solely on the build. All the labour I hired was on a day rate and I only ever employed people I could

The first-floor hall has spectacular views over the valley.

trust. Things went pretty smoothly but the small things that have gone wrong occurred on the days when I wasn't here! The family have been great – Ellie has done an enormous amount of work and my son Matthew did the wiring (under supervision) on his gap year.

"There is in fact a great deal of steel in the building. Under the entrance hall is a huge steel beam weighing 825kilos; it took ten people to lift it into place. The stone is Farmington limestone: the columns at the front were cut on site by the stonemason Richard Waite. He also cut a small window near the front door. It wasn't on the plans but one evening when he was building up the wall he happened to see a sunset from here and said to me 'David, you must have a window here to get these summer sunsets.' So he just went ahead and cut it in. I love to work this way with craftsmen, to give them their freedom and let the project reflect their individual tastes and skills.

"The turf roof is something of our own invention, however," says David. "We looked at doing a professional job but it was incredibly expensive, around £60/m$^2$, so instead we built up a structure using Silent Floor I-beams overlain with ply. Our waterproof layer is actually pond liner. It is not a turf roof as such but a collection of alpines, chosen by Ellie to be sun-proof and drought tolerant. It also has plastic pipe inlaid into it which we plan to use for solar water heating at a later date.

"The house is still unfinished but we thoroughly enjoy living in it, and we're very much looking forward to the challenges ahead!" says David.

## FACT FILE

**Names:** David and Ellie Austin
**Profession:** Architect and landscape designer
**Area:** Gloucestershire
**House type:** Three-bedroom masonry construction
**House size:** 290m$^2$
**Build route:** Selves with subcontractors
**Sap rating:** 100
**Mortgage:** Ecology BS
**Build time:** May '97-May '00
**Land cost:** £40,000
**Build cost:** £150,000

**Total cost:** £190,000
**House value:** £250,000
**Cost/m$^2$:** £520

**24%**
**COST SAVING**

## FLOORPLAN

**Two wings, one each side of the central area, house the bedrooms at ground floor level, while the main living spaces, including the kitchen, are situated on the first floor.**

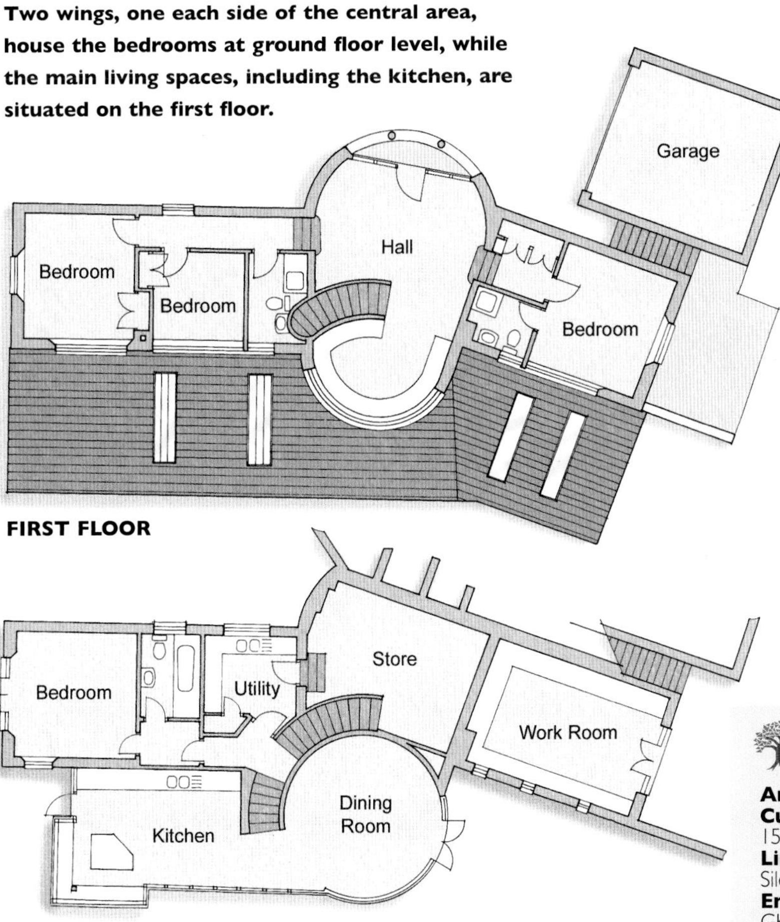


**FIRST FLOOR**

**GROUND FLOOR**

## USEFUL CONTACTS

**Architect** David Austin & Associates RIBA: 01453 836393; **Curved aluminium guttering** Dales Fabrications: 0115 930 1521; **Roof trusses** Oak Frame Carpentry: 01453 825092; **Limestone** Farmington Stone: 01451 860280; **Floor joists** SilentFloor from TrusJoist: www.trusjoist.com; **Structural Engineers** MD Hughes & Partners: 01453 824551; **Joinery** GK Joinery: 01453 885075.

# THE GRASS IS GREENER

Rory and Sylvie Robinson have succeeded in building a family-friendly eco home without blowing their budget.

WORDS: CAROLINE EDNIE PHOTOGRAPHY: ANDREW LEE

The turf was laid on top of four inches of earth and is designed to act as an insulator, retaining warmth in the house during winter and keeping the interior cool in summer.

Heat travels up into the open mezzanine area, keeping it warm. The flue also acts as a radiator.

After more than five years searching for a site, a further 18 months living in a caravan, and some heavy-duty hands-on work – including laying a natural turf roof – Rory Robinson proclaims that the new home that he and wife, Sylvie, now share with their three children, Natasha, Manon and Eric, is the ultimate family home. "I think your house should really reflect your philosophy of life, and this one certainly does," he says.

The result of all this patience – and toil – is Alltbeithe, a family eco home (although Rory prefers to refer to it as a "natural" home) set high upon the sloping site of an Inverness-shire woodland. Rory, who helped set up an organic food company when he first arrived in the area around 17 years ago, and Sylvie, a teacher in a local school, admit that this site had not been easy to come by. Indeed after five years of searching, it took a bit of luck to finally intervene before the couple managed to secure the site. "In 2003 the farmer that owned the surrounding land was selling off pockets, and almost sold the land to someone else, but they pulled out. When the farmer saw the land-seeking advertisement that we had placed in the local paper, he got in touch," explains Rory.

The beautiful hillside one-acre location, which provides the family with their own natural garden, also boasts stunning views. "It was a big attraction to us, and the children love it," says Rory.

In terms of the house itself, Rory and Sylvie originally explored the idea of constructing a standard timber frame kit house. "We thought this would be a cheaper option," explains Rory. "But the kit manufacturers we found weren't flexible enough for our needs. We had a post and beam construction in mind, and we also wanted to build an ecologically efficient house, so we eventually came to the conclusion that we needed an architect." The couple were particularly impressed with the eco homes of Inverness-based architect Neil Sutherland, so they approached him with a few ideas, and Neil agreed to take on the Robinson commission.

"We weren't too prescriptive when we went to Neil," continues Rory. "We had a list of no particular priority. We wanted to use as much local natural material and as many local tradesmen as possible. We also assumed materials would be low-toxicity. Points such as orientation were crucial – we wanted the bedrooms to face the east. We also wanted lots of glass overlooking the views of the loch. In addition, we wanted it to be very open plan and over two levels. First time round, Neil came up with plans that were more or less spot on, and the general layout changed very little from then on."

The house was constructed using an untreated Scots Douglas fir post and beam main frame, with spruce softwood panelling in between. The structure is set on a reclaimed drystone basecourse and the house is clad in vertical larch timber – a device that Rory believes reflects the neighbouring trees. Another respectful nod to the surrounding native woodland landscape is the natural turf roof, which appears almost like a wild meadow in its own right – although the solar panels do give away its function. The house is super insulated using Warmcel, which also fits the environmental bill and allows the building to utilise breathing wall technology. Solar energy contributes to the ➤

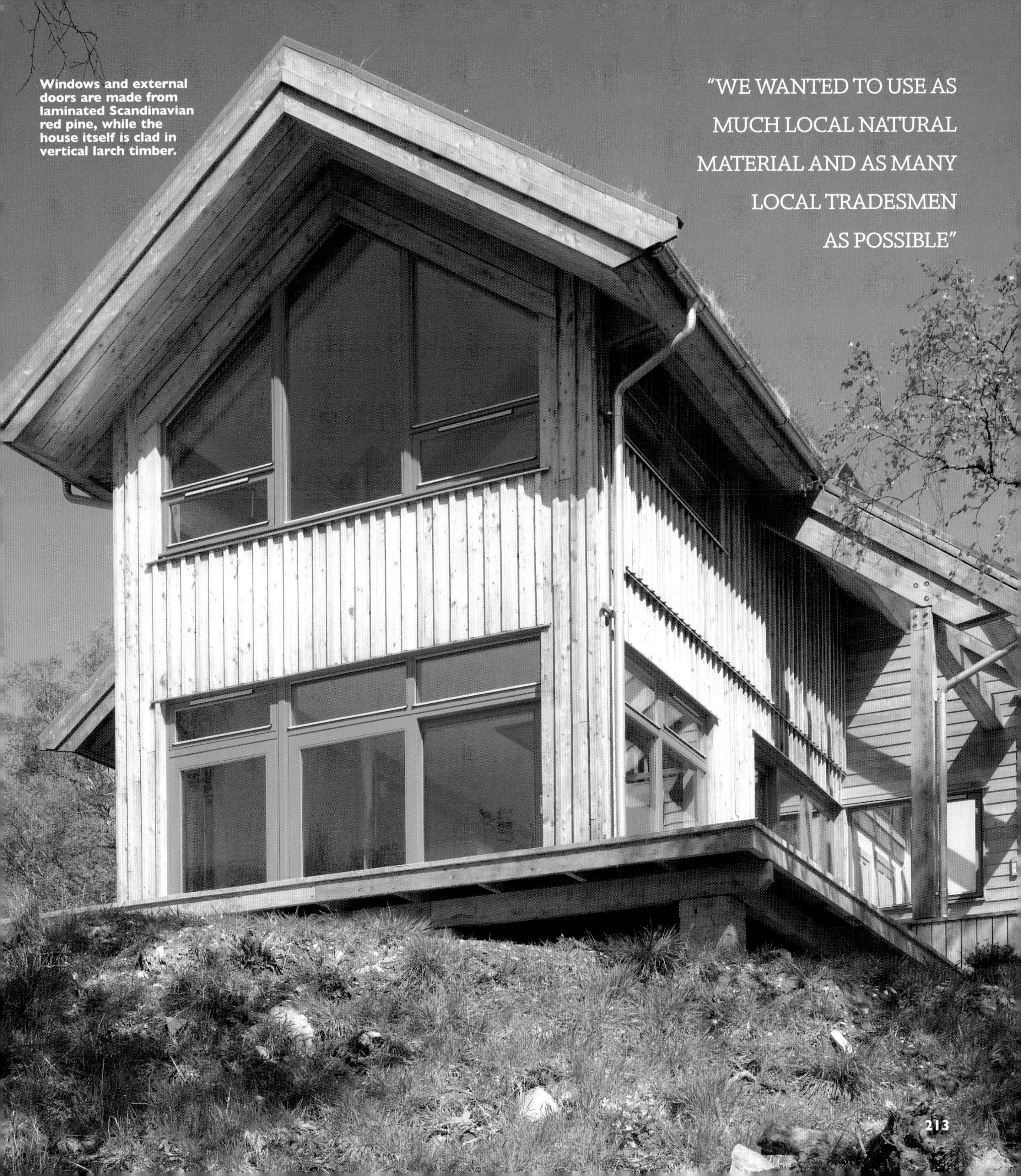

Windows and external doors are made from laminated Scandinavian red pine, while the house itself is clad in vertical larch timber.

"WE WANTED TO USE AS MUCH LOCAL NATURAL MATERIAL AND AS MANY LOCAL TRADESMEN AS POSSIBLE"

**Right: The linoleum floor in the kitchen is both practical and ecologically friendly: it's biodegradable and constructed from entirely natural materials.**
**Above: Solar energy heats the house passively through extensive glazing to the south and west. This is stored in the high thermal mass in the floor slab.**

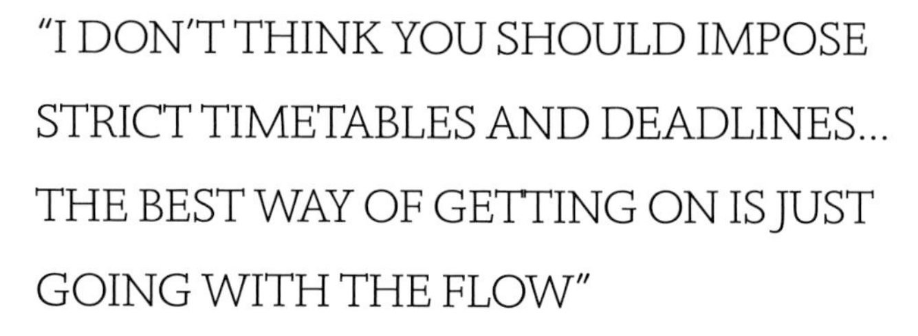

"I DON'T THINK YOU SHOULD IMPOSE STRICT TIMETABLES AND DEADLINES... THE BEST WAY OF GETTING ON IS JUST GOING WITH THE FLOW"

heating of the house passively through the extensive glazing to the south and west. This is then stored in the high thermal mass in the floor slab, and also passively in the sun space to the entrance of the house, which opens to the upper floor. Finally, energy is stored actively in the solar panels on the roof, which connect to the domestic hot water tank, and heat everything apart from the underfloor heating.

The house incorporates one level change on the ground floor, which responds to the sloping site. This also forms an effective contrast to the more cellular bedroom and bathroom spaces, which are situated to the east of the house. Externally, a series of timber decks from the principal living spaces further extend these rooms. Two of the children's bedrooms are located downstairs with the master bedroom and bathroom located above on the mezzanine level. Apart from these rooms, the mezzanine has been left open to create a greater sense of space and take advantage of the views.

"Upstairs on the open mezzanine it's lovely and warm because heat travels up. The glazing at the main entrance canopy also adds to this warmth," says Rory. In addition to the solar gain, the Robinsons have underfloor heating on the ground level. "If I'd been very eco I'd have used a non-fossil fuel, such as a renewable light wood boiler, but I've used wood before and it really is hard work. So we went for oil, and we've only used one tank over the five winter months. The underfloor heating works really well – it makes the living area so comfortable, and we do spend a lot of time in this part of the house. Our family time centres round preparing and eating our meals, so the kitchen/dining/living area is the heart of the house.

"We see this as our long-term family home, which is why we spent extra on many good-quality finishes: we want to benefit from them for the next 20 years. We took the long view," says Rory. This strategy means that Alltbeithe boasts laminated Scandinavian red pine used by local manufacturer Treecraft for windows and external doors; French oak and Scandinavian Scot's pine floorboards to the upper floor; Scandinavian dressed redwood finishes for internal cills and architrave; a Scottish oak

staircase made by local craftsman Adrian Ellis, and linoleum in the living area.

Rory admits, however, that in order to achieve this quality of structural and material finishes, some cost-cutting did have to take place to meet the budget's parameters. The kitchen, for example, is off the shelf from MFI. By literally mucking in during construction, the Robinsons also managed to keep costs down. Last summer, after the main frame had been constructed, Rory and Sylvie took on the task of laying the turf roof themselves with the help of a friend, and also had a turf party, where a group of friends came up to the site one Saturday last summer to help the couple with the lifting, carrying and laying of the turf. "Laying it is just like a jigsaw puzzle. I haven't had to maintain it, but our site isn't that exposed to the elements. It's four inches of earth and, for its thickness, it's not a good insulator, but on the other hand it reduces temperature changes – so on a very hot day it won't overcook the house. Primarily it's aesthetic."

The Robinsons also did their fair share of site supervision. "Although the architect supervised at the really important stages, we found ourselves being clerks of work the rest of the time," says Rory.

Although the family are installed full-time in Alltbeithe, Rory admits there are a few finishing touches here and there that need to be done, but makes no attempt to implement a strict timetable. "I don't think you should impose strict timetables and deadlines in rural areas anyway. It's a small community and I think the best way of getting on is just going with the flow."

## FACT FILE

**Names:** Rory and Sylvie Robinson
**Professions:** Owner of organic food company and teacher
**Area:** Balnain, Highlands
**House type:** Detached, four-bedroom family house
**House size:** 135m$^2$
**Build route:** Self-managed plus architect
**Finance:** Private
**Construction:** Douglas fir post and beam frame, with untreated spruce softwood panelling in between
**Build time:** 18 months

**Land cost:** £35,000
**Build cost:** £115,000
**Total cost:** £150,000
**Current value:** Unknown
**Cost/m$^2$:** £851

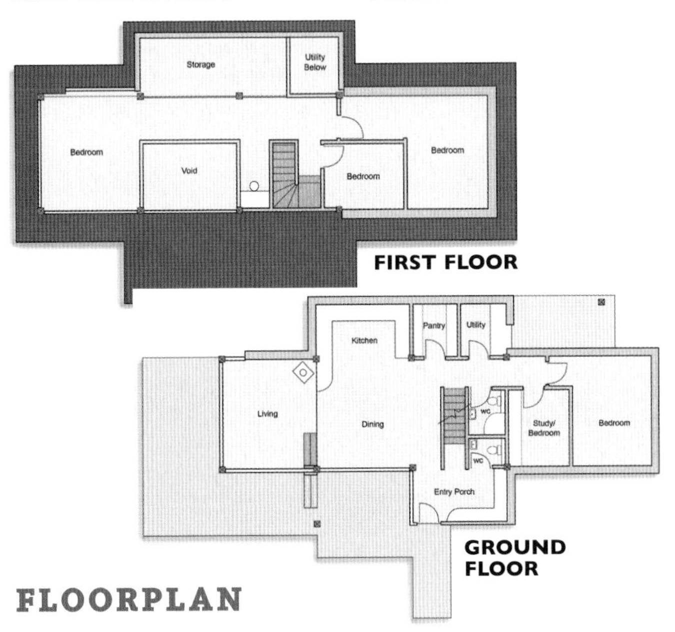


## FLOORPLAN

**An open-plan kitchen dining area – double height – is the focal point of the ground floor, which also houses a bedroom in addition to the three upstairs.**

## USEFUL CONTACTS

**Architect** Neil Sutherland Architects: 01463 709993 **Structural engineer** AF Crudens: 01463 719200 **Main contractor** Kenny Beaton & John Dalgetty: 01463 811327 **Joinery** The Stornoway Trust: 01851 706916 **Cabinet maker, timber supplier, staircase, balustrade and decking** Adrian Ellis: 01456 476268 **Stonemason** Tom Nelson: 01456 450506

# GO GREEN

WELCOME TO *GREEN HOMES'* QUICK REFERENCE GUIDE TO LOW-ENERGY, ECO-FRIENDLY BUILDING METHODS, PRODUCTS AND BEST PRACTICE ADVICE

**Large, south-facing windows will bring in lots of light, but to avoid over-heating, plant trees in front to provide shade.**

# A BASIC ECO MAKEOVER

**Just because your home is old no longer means it has to use up lots of energy. From extra insulation to energy generation, you can join the growing army of renovators carrying out an 'eco makeover'.**

Owning an old house has long been an excuse for using up lots of energy – and money – in keeping it light and warm. Not any longer. More and more renovators are taking advantage of a growing range of features that can be installed just as well into an existing house as they can be a new one — meaning that the traditional benefits of owning a new home, such as greater warmth and comfort – and lower energy bills – are now available to all. If you know what features make environmental as well as financial sense, modest investment at the planning stage of a renovation project can transform the way you enjoy the house for many years to come.

Eco makeovers, as they are known, range from simply adding extra insulation into the existing structure to generating energy on site (usually through solar or wind power). They are also very much in the news thanks in part to higher prices, and also the fact that some pretty high-profile renovators are considering them. Perhaps the most well-known house on the receiving end of an eco makeover is the one in West London belonging to the new Conservative leader David Cameron. The redesign is being handled by architect Alex Michaelis (0207 221 1237), with a brief to make it perform as ecologically as any large brick Edwardian house can. "In a sense, David Cameron's house is no different from any large five-bedroom unlisted house of its generation," says Alex. On the following pages we examine, with Alex's advice, the best options.

## INSULATION: WALLS AND LOFT

Around 40% of the heat that is lost from a home is lost through walls and the roof (the rest is lost through the floor and openings). This figure can be significantly reduced through the installation of simple loft insulation, which will cost between £130-200 to install and save around £150 a year in energy bills. Cavity wall insulation costs from around £135 to install and can save around £100 a year, while other forms of internal and external wall insulation cost a little more and have slightly longer payback times. Grants are available to further increase the financial incentive to undertake this type of project (see www.est.org.uk). "Insulation takes many forms," says Alex Michaelis. "In my own house I have used insulation on both the inside and outside of my solid walls. With David Cameron's house, because it will be a retrofit, I shall look at the possibility of adding insulation to the exterior of the walls. This might be possible on the back as it is in a terrace. At the

**ECO FACT**

**£271 MILLION AND 2.5 MILLION TONS OF CARBON DIOXIDE ARE BEING WASTED EVERY YEAR BECAUSE 40% OF THE UK'S HOUSING STOCK DOES NOT HAVE CAVITY WALL INSULATION**

WORDS: CLIVE FEWINS

ECO FACT
**27% OF THE UK'S CARBON DIOXIDE EMISSIONS ARE ACCOUNTED FOR BY PRIVATE HOUSING.**

front it is most unlikely because it will almost certainly be necessary to preserve the façade to keep the unity of the street. Therefore I shall probably look at ways of insulating the front of the house from the inside. Because the house has large sliding sash windows, they too will need to be insulated. Windows such as this can be insulated using brush seals as a means of achieving further air-tightness." Most renovators, however, will want to install double glazing to reduce heat loss through windows, although timber is preferential from an environmental perspective to uPVC.

## OTHER BASIC MEASURES

Lighting accounts for some 15% of an electricity bill. As each energy-saving light bulb saves some £7 a year, the benefits are potentially significant. On an even simpler level, ensuring that the new appliances you install are approved by the EST (look for the logo) means that, for example, on a fridge, you can reduce its energy consumption by two thirds – saving you £35 a year. A high-efficiency condensing boiler, particularly one that uses weather compensation control, will save you around £180 a year, while simply adding a jacket and lagging to the hot water cylinder and pipes will cover its outlay in a year. In particularly dark areas of the upper storeys of homes, a lightpipe might be an effective alternative to constant artificial lighting – light is captured from the roof and channelled down a mirrored tube. They cost between £200-700, depending on size and distance from roof.

## ENERGY GENERATION

PV cells are perhaps the best known form of renewable energy and operate by turning solar radiation into electricity. The average UK home could generate around half of its own electricity through this method, providing it is used efficiently and the home has a large south-facing roof. The greater the intensity of the sunlight, the greater the amount of electricity produced. The grant situation is currently confused by the closure of the Clear Skies scheme in England and Wales, although in Scotland and Northern Ireland, grants of up to 50% are still available. Payback times on PV systems vary but are usually considered to be at least 10 years. Straightforward solar panels, which simply use the sun to heat water, tend to offer a much shorter payback period but do not enjoy the same grant status. Alex Michaelis is a great advocate of solar panels, particularly the evacuated tube type of system. "With straightforward solar thermal panels of this sort, I find that for a few hundred pounds you can easily get your money back in a few years because of the energy saved in heating your hot water," he says.

## AIR CIRCULATION SYSTEMS

The more substantial the structural elements of the renovation project, the more possible it is to install new, more energy-efficient central ventilation systems with heat recovery. The key to such systems is a heat exchanger, usually installed in the roof. This unit draws air from the moist areas of the house – the kitchen and bathrooms – and expels it through the heat exchanger. At the same time cold air from the outside is drawn in and warmed by the heat in the outgoing moist air. This warm air is then transferred to the bedrooms and living areas. The best heat exchange units claim to be able to recover about 90% of the heat that would otherwise be deposited outside. Installing all the ducting necessary to make systems like this work can be tricky and messy when the job is a retrofit. However,

when this takes place in a period house that is having a total refurbishment, it is unlikely to cause more disturbance than there is already.

## REDUCING WATER CONSUMPTION

Although water bills are still relatively cheap in the UK, drought and water shortages have brought the focus back on to the way domestic properties use water – they currently account for up to 1,000 litres a day. All new builds are now required to install a water meter rather than pay a flat rate charge.

Renovators interested in cutting their consumption would do well to follow this measure, as well as installing low-flush WCs and simple showers, rather than power showers, which use around 20 litres of water a minute. Even when compared to a bath which uses around 100 litres in total, power showers look very high in usage and those interested in conserving water should consider avoiding them. Fitting flow restrictors might be an alternative.

ECO FACT

**TO MEET THE WORLD'S DEMAND FOR TIMBER, 20 FOOTBALL PITCHES OF FOREST ARE CLEARED EVERY MINUTE**

For those contemplating a more significant renovation scheme, the installation of a rainwater harvesting system should be considered, as it can potentially reduce a household's water consumption by around half. Rainwater harvesting systems, which cost around £3-3,500 installed, recycle rainwater and use it to flush toilets, water gardens and for the washing machine (www.freerain.co.uk).

## WIND GENERATORS

Using the wind to produce energy is becoming a viable option for individual dwellings. The Swift Rooftop Wind Energy System is just 1.6m in diameter and costs around £1,300. It generates around 1.5kWh of energy every time it turns. An alternative is the Windsave, which operates from the ground and requires a pole, although the manufacturers claim the £995 (plus 5% VAT) system can save a third of your annual electricity costs, giving a payback time of around five years.

## GROUND SOURCE HEAT PUMPS

The temperature at around 5-10m underneath the ground surface is at a constant level of around 10°C. This heat can be extracted using the same technology as is used in domestic fridges to provide up to four units of energy for every one unit of electricity used to power the system. Installation costs are between £8,000-12,000, which is reduced somewhat under grants available on the new Low Carbon Buildings Programme. Most experts give the payback time at around ten years, and for renovators installation of a heat pump is only really practical if a significant amount of renovation work is taking place. ●

Case Study No1: Donna Grey's 1960s house before (right) and after the renovation.

Case Study No2: Karen used wood from the trees in her garden for the stairs and kitchen.

# REMODELLING WITH ECO MATERIALS

Donna Grey and Karen Hughs both restyled their houses using eco-friendly and recyled materials.

## CASE STUDY NO1

While renovating and remodelling her tired 1960s home, Donna Gray took the opportunity to incorporate a whole range of low-energy features.

Not only did Donna, a freelance designer, want to turn her home into a showcase for some of her designs, but she also wanted the overall structure to be made energy efficient and to incorporate sustainable, eco-friendly materials.

Donna commissioned architect Duncan Baker-Brown for the project after being impressed by his approach to low-energy sustainable architecture.

Donna has chosen eco-friendly materials wherever possible, including non-toxic paints and Homatherm cellulose thermal insulation, made from recycled paper and jute sacking, which is treated with borax to make it resistant to decomposition and fire. This insulation also absorbs and diffuses moisture and therefore moderates humidity as well as temperature variation. A solar panel on the roof now warms water for the underfloor heating pipes, with a boiler providing back-up should it be required.

## CASE STUDY NO2

Karen Hughes stumbled upon her oak-framed house, which had lain empty for 17 years, while walking in the woods. When she bought the property there was no running water, electricity or proper road access.

Eight years later and Karen has managed to retain all the character of her cottage – despite being advised by the builders to knock it down and start again.

She had added numerous ecologically friendly features, as well as a new bedroom and a small extension for allow room for a new kitchen and bathroom.

Karen was keen that the cottage would be able to 'breathe' – she "didn't want to use materials that would end up getting into the soil." Instead, she relied on natural, organic or recycled materials such as limewash and earth pigments.

An Italian cedar from the garden was felled and reused for the stairs, kitchen worktops and interior window ledges, while another tree from the garden was planked up and used as flooring in the kitchen, the upstairs landing and as a surround for the bath.

# ECO HOUSE DESIGN

How can a house design be inherently eco-friendly? Let's have a look at the basics of eco house design and explains where to start

Designing an ecologically friendly home is a balancing act and cost will always be a major factor. It is an unfortunate truth that building projects tend to exceed their budgets – and green homes are no exception. Sustainable materials still tend to be more expensive than their alternatives and skilled people are hard to find and come at a premium. So trim the design, not the build. Cutting costs at the end of the project can result in a building that is a compromise, which satisfies no one.

## SHAPE AND ORIENTATION

The shape, size, location, topography and surroundings of your plot will all influence what can and can't be done. South-facing elevations and roof planes are great, and tall, thin windows on the south will introduce more light and heat to the back of a room. But south-facing windows can overheat the home in summer. Planting deciduous trees in front of the window will let sunlight in during winter and provide dappled shade in summer.

Designing an eco home is not just about the building. It is also about how the building sits in its landscape. Minimise its impact and think about its surroundings. Permeable tarmac for the drive, for instance, will reduce run-off and put water back into the land. Trees, shrubs and 'wild' areas will increase biodiversity and interest.

If you're using solar-heat gains you will need some masonry mass to hold the heat. A concrete slab with tile covering in front of a south-facing window will hold heat well, but there is not much benefit in putting that opposite north-facing windows. Put the mass where it will do some good and minimise cement use.

## MATERIALS

Think first of using recycled or salvaged materials: they have virtually zero embodied energy, are available locally and cost less than a new equivalent. Timber, roof slates, and even bricks are all readily available and the range of new goods from recycled materials is growing – from carpets to roof tiles.

As to structure, timber frame, especially sustainable timber, with sustainable insulation, has true eco credentials. The rules are: timber first, steel last, less cement and avoid PVCu.

## ENERGY

Let's be clear, the greatest environmental impact of a house is from the fossil fuels it burns for its energy. No amount of eco-certified bamboo flooring can compensate for the impact of an oil- or gas-guzzling house. Direct the budget to energy conservation first, energy generation second and everything else last.

## OPERATION

Eco design is as much about living in an eco-friendly way as about the house itself. Minimising your energy use, providing space for recycling bins or clothes drying, and using grey water will all help to reduce your impact on the planet, and the impact of the people that come after you.

## COST

In the long term, a well-designed eco house will recover its extra expense in lower running costs. It will command a premium price and will be more attractive to the buyer than the estate home down the road. More than ever before, a sustainable home is a sound investment. ●

## DID YOU KNOW

**● Thermo-hemp from Ecological Building Systems is a great natural insulation material. The price and performance are comparable to glass-fibre quilt, and it is available in mat and roll form.**

**● Grey-water recycling is a great idea but the tank can cost £1,500 plus. Try a local scrap metal merchant and buy a couple of second-hand copper hot-water cylinders. They will cost £10 to £20 each and with a bit of pipework will do a good job, with zero embodied energy.**

**● According to an Energy Saving Trust survey, the average 8m skip leaving a building site contains £1,300 worth of materials. Sort your waste, save the stuff that has value and re-use it.**

**● To increase the thermal qualities of solid walls when renovating, try Sempatap from MGC Ltd (www.mgcltd.co.uk), a high-grade latex form bonded to a tough non-woven glass fibre substrate. As easy to fix as wallpaper, it will increase the insulation level by about 20%.**

WORDS: TIM PULLEN

**You can help rainwater return more quickly to source by choosing suitable landscaping materials – such as log paving or gravel – and including natural ponds that act as a store for heavy rainfall.**

# ECO FRIENDLY DRAINAGE

In this era of extreme weather, every eco-friendly house should have an eco-friendly way of disposing of water.

Although it's often known as Sustainable Urban Drainage Systems (SUDS), the principle of rainwater disposal which does not put undue pressure on existing resources and infrastructure is applicable to any site, rural or urban. In general, it is best to deal with rainwater discharge locally, returning the water to the natural drainage system as near to the source as possible.

Most experts are agreed that one of the imminent major impacts of climate change – some would say that it is happening already – is an increase in 'extreme weather events' such as storms and floods. Periods of unusually heavy rainfall may result in antiquated or inadequate drainage systems unable to take the increased flow and 'backing up', causing local flooding and pollution. Surface water runoff can contain contaminants such as oil, organic matter and toxic metals. Although often at low levels, cumulatively they can result in poor water quality in rivers and groundwater, affecting biodiversity, amenity value and potential water abstraction. After heavy rain, the first flush is often highly polluting.

If we can slow down the rate of surface water runoff and provide temporary storage or 'attenuation' for rainwater on site, this will ease the pressure on existing drains. Minimising runoff can be achieved at the design stage by using 'green' or vegetated roofs, which will naturally filter and slow the rate of discharge. Porous surfaces, such as paving, gravel or grass, rather than seamless hard concrete or tarmac, will allow rainwater to percolate into the ground, eventually refilling underground aquifers which are much more important than reservoirs for our long-term water reserves. This will also reduce the flooding risk both on site and downstream.

Filter strips and swales are grass banks and ditches which mimic natural drainage patterns by slowing and filtering the flow and removing pollutants. They provide temporary storage and can be integrated into landscaped areas and road verges. Basins and ponds can provide more permanent storage and should be designed with extra capacity to enable storage of flood water. Both increase areas of wetland habitat, important for biodiversity and providing educational and leisure amenities. ●

## GOOD DRAINAGE PRACTICE

- **Rainwater harvesting and reuse (e.g. water butts or tanks)**
- **Green roof on building**
- **Retention of 'soft' surfaces (grass); specification of porous hard surfaces; avoid concrete and tarmac**

USEFUL CONTACTS: The Environment Agency: www.ciria.org/suds/suds_techniques.htm

## DID YOU KNOW

- **Prince Charles has submitted plans to rebuild a mansion in Herefordshire as an eco home for son William. The new £5m home will feature solar panels, biomass boilers and sheep's wool insulation.**
- **A new rainwater harvesting system which is designed for use in the garden has been launched by Klargester. Known as Raintrap, the system comprises a filter, an underground storage tank and a pump. Rainwater runs down the roof and into the guttering and downpipes and passes through a filter. The water is stored in an underground tank available in three sizes. You can then access the water at a constant pressure. From around £1,000. www.klargester.com**
- **The Government has unveiled plans to encourage green homebuilding. New regulations, launched in stages over the next ten years, will ensure all homes are carbon neutral by 2016. Around 27 per cent of all carbon emissions in the UK are from energy used to heat, light and run homes.**

WORDS: CINDY HARRIS AND PAT BORER

# EARTH SHELTERED BUILDINGS

## Partially burying your house into the ground can prevent excessive heat loss from the building.

Often confused with buildings made from earth, earth sheltered building is in fact a construction method which partially buries the building into the ground. It is often carried out on sloping sites so that the building is virtually invisible from the uphill side, and the earth is 'bermed' to cover the roof and three walls. The fourth wall, ideally facing south, is usually highly glazed to allow in maximum daylight and solar energy.

Earth sheltered buildings are not made from earth. They have to be constructed with materials that are impervious to damp, and are usually made from reinforced concrete. These are heavyweight buildings, able to withstand the relatively high loads of dense, wet earth. The earth covered elements of the building (roof, rear and side walls) are protected by a tough, durable, waterproof membrane which should stop moisture migrating through to the inside. Extra drainage may be installed around the building footprint.

The earth covering is often claimed to be an extra insulation layer, but in fact earth is not a particularly good insulator. It is, however, a high thermal mass material and, once past the frost line, earth will remain at a constant temperature of about 11-12°C throughout the year in the UK. This prevents excessive heat loss from the building. The earth also acts as a heat sink and store, absorbing excess heat in hot weather and releasing heat in cold weather. However, for this exchange of heat to work well, the building's walls and floors have to be well coupled to the surrounding earth, and not insulated. The roof should always be well insulated, as with any 'green' roof.

Many environmental benefits are claimed for earth sheltered houses. While it is true that living underground offers more comfortable temperatures – as our cave-dwelling ancestors discovered – most earth sheltered buildings today will require some form of external heating, albeit for a reduced heating demand. If earth sheltering is combined with a passive solar design on the south(ish) facing elevation, then solar energy can be maximised, stored and re-used.

This construction method offers excellent insulation from external noise, and requires little external maintenance. Its main disadvantage is the inevitable lack of natural daylight in the main body of the house, as three out of the four walls will have no windows, although this can be partly compensated for by the use of lightpipes. ●

**HOW DO THEY WORK?**

**Earth sheltered buildings are not made from earth – they are sheltered by it. They are in fact constructed from heavyweight waterproof materials. The earth helps keep the building warm.**

USEFUL CONTACTS: The British Earth Sheltering Association: www.besa-uk.org; Earth Homes: 07968 951215 www.earthhomes.co.uk; Polarwall: 01392 841777 www.polarwall.co.uk

## DID YOU KNOW

- **Buyers are prepared to pay more for eco homes, according to recent research by self-build company GreenerLiving Homes, who predicts that a well-designed, desirable and eco-friendly house could add up to a ten per cent premium on its resale value compared to a similarly sized, but unsustainable, property.**
- **A new website has been launched offering a range of eco-friendly products for sale, as well as free advice on how to incorporate more green features into your home. Log on to www.greenandeasy.co.uk to find out what you can do to ease your eco conscience.**
- **The world's first fully integrated solar roofing system has been launched by Nu-lok Roofing Systems. The ingenious invention simply locks the panels into the roof's battens to provide a safe, green power source, which seamlessly integrates into the existing roof. Tel: 01895 622689 www.nu-lok.com**

# THE BIG ECO HOMES TEST

## How will the new Home Information Pack Energy Performance Certificate affect your home?

As part of the European 'Energy Performance in Buildings' Directive, which was adopted by the UK Government in 2003, much clearer information will be made available to the general public on how much energy buildings consume, and how energy efficient they are. This Energy Performance Certificate (EPC) will be included in the new Home Information Packs (HIPs), to be launched on 1st June 2007. This information will need to be produced for all newly built houses, or any house offered for sale.

In spite of what its critics may say, the idea behind this is not to increase the hassle of property conveyancing, nor to line the pockets of estate agents. The point is to increase public awareness of an aspect of buildings which until now has been largely hidden. We know whether or not we like the appearance of a house, its size, layout and location; now we are also being offered information on its likely energy consumption and fuel bills. Not only that, but the certificate will also contain information on potential energy-efficiency improvements, their costs and benefits. It is hoped that this will increase public demand for energy-efficient buildings and encourage homeowners and builders to invest in low-carbon measures.

So what will all this mean for people buying and selling houses? Vendors, through their estate agents, will need to employ a competent person to produce a certificate which rates the energy performance of the house from A-G, very much like the ratings you see on fridges or washing machines. The EPCs will outline the financial costs and carbon emissions of heating, hot water and lighting, and give practical advice on how to cut costs and reduce emissions.

The certificate will then be included in the HIP, which will also include searches and other legal documents. The whole system is designed to tackle the uncertainty and lack of transparency in house purchases, leading hopefully to a reduced number of failed transactions. ●

### WHAT WILL IT MEAN FOR YOU

- **Your house will be rated on a scale of A-G.**
- **Your certificate will outline the costs and carbon emissions of heating, hot water and lighting.**
- **You will be given advice on how to reduce emissions.**

### DID YOU KNOW

- **If you fancy it, you can train to become a Home Condition Inspector and issuer of Energy Performance Certificates yourself. See www.bre.co.uk for details.**
- **The Biomass Energy Centre has launched a website to provide a portal for technical information, grant availability and best practice guidance. See www.biomassenergycentre.org.uk.**
- **The DTI has launched a new website that shows all renewable energy projects across the UK via an interactive map. All significant projects – approved and built – are included, together with their generating capacity. http://maps.restats.org.uk**
- **The Centre for Alternative Technology has released a series of books for people who want to tackle global climate change at home. Titles include Choosing Windpower and Solar Water Heating. Log onto www.cat.org.uk to buy.**

WORDS: CINDY HARRIS AND PAT BORER

**Windows in more than one wall will give an increased and even distribution of daylight, avoid glare, and add to the sense of well-being in internal spaces.**

# HEALTHY HOMES

## Cut down indoor pollutants to reduce allergic reactions and immune system deficiencies.

All forms of shelter were historically developed to improve our physical and mental well-being. At its most basic, this means protecting ourselves from extremes of weather, and it is still true that cold, damp homes can affect people's health, leading to asthma and respiratory and cardiovascular diseases. So the first and foremost requirement for a healthy home is to ensure that it has a warm, even temperature, by maximising insulation and draught-proofing, and controlling ventilation.

Introducing generous amounts of daylight into the home, via several large windows or rooflights, will lift the spirits, provide visual comfort and reduce eye strain. External noise and pollution can be reduced by the installation of double glazing and by placing ventilators or extractor fans away from busy roads.

Paradoxically, our desire for greater comfort and convenience has led to less healthy internal environments, as indoor pollutants released from an increasing number of synthetic materials have accumulated in more airtight buildings. Historically, houses that were draughty and had open chimneys could easily vent away any toxic products. Of course, we would not want to return to the days of uncomfortable, hard-to-heat homes, but we should exercise caution in the furnishings and fittings, and floor and wall finishes that we introduce into our homes.

One of the major types of indoor pollutants is Volatile Organic Compounds (VOCs), a highly reactive class of chemicals emitting a gas. VOCs are emitted from a range of construction and household products including:

- Paints, stains and glues, even water-based versions
- Timber treatments and preservatives
- Isocyanurate foam in foam fillers and adhesives

VOCs can produce allergic responses and lead to immune system deficiencies. They can be minimised by using low-VOC paints (read the small print on the tin) and by avoiding synthetic surfaces such as vinyl floors and walls. Timber sheet products such as chipboard and MDF contain high proportions of VOC-based glues.

In general, natural products are always best, such as timber floors, wool carpets and cotton soft furnishings. Any known toxic materials, such as lead or asbestos, should be professionally removed. ●

### WAYS TO A HEALTHY HOME

- **Large amounts of daylight**
- **Maximise insulation and draught-proofing, and control ventilation**
- **Do not use VOC-based products**
- **Opt for natural products, such as wood**

This new house in Dorset has a solar panel to assist with heating the hot water, supplied by Solar Sense (01792 371690).

# WHICH FORM OF SOLAR POWER?

The sun is a cheap and easy source of alternative energy and one we can tap into in two ways...

There are two types of 'solar panels' that use the sun's energy to help service our buildings. They are very different from each other and should not be confused.

Solar thermal panels provide heat, usually for domestic hot water. Solar photovoltaics provide electricity. We are all familiar with this technology on a small scale, in the form of solar powered calculators or roadside signs.

## HOW DO THEY WORK?

Photovoltaics use semi-conductor cells which produce a current when light hits them. The cells are arranged in panels which typically produce 10-100 watts of electricity in bright sunlight.

Over the last 50 years the efficiency of PVs has improved significantly and costs have come down. The major manufacturers now guarantee the output of their panels for 20 years.

Ideally panels should be mounted at an angle of 10-35 from the horizontal, pointing south (or south-east or south-west), and be unshaded. They can form part of the roof covering and so save on slates/tiles. Depending on the type, 7-10m² of panels will produce 1kW of electricity in bright sunlight, or about 25% of an average house's annual electricity requirement.

## WHAT ARE THE COSTS?

A 2.5kW system will cost around £13,000 installed, and produce 1,800kW hours of electricity each year. Grants are available from the Energy Saving Trust for £2,500/kW, in this case £6,250.

## WHAT ARE THE BENEFITS?

Photovoltaic panels will convert your house into a miniature power station. When it is sunny you will in all likelihood probably produce more electricity than you will use, which means surplus electricity will be exported to the National Grid.

Savings are likely to be around £150-£200 per year for the £13,000 system.

Without the grants taken into account, this gives a payback period of some 65 years – which for many will mean that the benefits have to be more than purely financial.

The system will also save around 750kg of carbon dioxide emissions each year. ●

USEFUL CONTACTS: Energy Saving Trust: www.est.org.uk/solar; British Photovoltaic Assoc: www.pv-uk.org.uk; Energy Equipment Testing Serv: www.est.co.uk; National Energy Foundation: www.greenenergy.org.uk; Centre for Alternative Tech: www.cat.org.uk; Viessmann: 01952 675000

WORDS: CINDY HARRIS

Heavy and dense materials such as stone absorb the heat gained during the day and release it during the night.

Fixed louvres and sun 'fins' should be used to reduce the risk of a house overheating owing to south-facing glazing.

# PASSIVE SOLAR DESIGN

**Passive solar design can harness free energy from the sun to help heat your home, reducing $CO^2$ emissions and household bills.**

Passive solar design (PSD)uses the sun's heat and light as free, renewable energy to create more pleasant places to live in and to reduce fossil fuel use and bills. 'Passive' in this context means that the transfer of heat operates automatically, by virtue of the way the building is designed, rather than relying on 'active' systems which use pumps or fans.

PSD is based around the main principles of good environmental design, such as high levels of insulation, air tightness, adequate ventilation, efficient glazing and heating systems, though there are additional design considerations.

## ORIENTATION AND LAYOUT

Orientation refers to the direction which the main façade of the building 'faces', and the main façade is the one behind which the main living areas are situated. So when we say a building faces south, we mean that the internal layout is arranged so that the living room, kitchen, study or main bedroom, have windows which face south. This is the optimum orientation for capturing solar energy, particularly in winter.

## FENESTRATION

This is a useful architectural term, which simply means the size and location of windows (if you remember your GCSE French it will make more sense!). Along with a southern orientation, it is important to arrange areas of glazing, such as windows and patio doors, so that they predominate on the south side and are minimised on the north side. Glazing on east and west façades should be somewhere in between and reflect the activities likely to take place in the rooms behind, for example an east-facing breakfast area will get the morning sun; a living room or relaxing space would benefit from the western evening sun.

## SHADING

There is always the danger of overheating if your house gets too much sun – this is particularly noticeable in conservatories. So consider building in shading devices, such as overhanging eaves, which will keep out the high midday summer sun, fixed louvers, blinds which can be lowered as required, and plenty of opening windows. Highly glazed west-facing rooms are prone to over heating as they will get the setting summer sun at the end of a hot day. It is very important that using PSD does not lead to an increase in air conditioning.

## THERMAL MASS

Having arranged the building to capture lots of solar energy during the day, it is useful if we can store that energy until the evening, and make use of it then. This is exactly what thermal mass enables us to do, using materials such as stone, concrete, ceramic tiles, compacted earth etc. Basically, anything heavy and dense will absorb excess heat from the air and release that heat as temperatures drop below the temperature of the material. If, for example, a stone or tiled floor in front of large south-facing windows is also dark in colour, that will increase its absorption of solar energy.
One word of warning though: for your high mass floor or wall to capture and store solar heat, it needs to have direct access to sunshine. That means no carpets or curtains getting in the way. It also needs to be well insulated below. ●

WORDS: CINDY HARRIS AND PAT BORER

# GROUND SOURCE HEAT PUMPS

## How to use heat from the ground around your house.

In this country, the average temperature 3m below ground is always around 10°C and a ground source heat pump (GSHP) can be used to transfer this heat to the house.

It's almost too good to be true – a heating system that sucks heat out the ground, that's clean, cheap to run, easily controlled, has low $CO^2$ emissions – and you can get a Government grant for some of the cost. Well, it's mostly true, but there are a lot of 'ifs'.

A heat pump works in a similar way to a domestic refrigerator – an electric pump compresses and moves a refrigerant gas in a sealed circuit that can extract heat from a low-temperature source (from pipes in the ground below your garden or in a borehole) and upgrade it to a higher temperature (to the radiators in your house). Although GSHPs use renewable solar energy from the ground, which is continually topped-up by the sun all year round, we need to use electricity – mostly from fossil fuels – to extract it.

The efficiency, or Coefficient of Performance (COP), of the system and hence its cost effectiveness depends on the temperatures of the two halves of the system – in the ground and in your house – being as close together as possible. So here is the first 'if': If you have a garden large enough to accommodate 100m-150m of trench 2m deep, or can have one or more 50m deep boreholes drilled, and if you can install underfloor heating or double-size radiators that can deliver heat at low temperatures (35–50°C), then your system may achieve a reasonable COP of 3.5 (3.5 units of heat given out for every unit of electricity) – considered the break-even point.

There is another 'if'. A large heat pump may require a three-phase electrical supply to run it – large heat pumps are also costly and require large, extensive heat sources. To keep the whole system and the capital costs to a minimum, your highest priority is to insulate and draught-strip the house as well as you possibly can. The heat pump then has an easier job to do, as the radiator or underfloor water temperature will be lower in a low-energy building. ●

**IN ACTION**

**This new 210m² house in West Wales (top) is very well insulated (walls 200mm sheep's wool; roof 300mm cellulose fill; Low-E windows) and has underfloor heating in the insulated ground floor slab. The installation, by John Cantor Heat Pumps Ltd, consists of 600m of pipe laid in a 150m trench to a depth of 2m (above left) connected to a 7kW single phase Stiebel Eltron heat pump (above right). This in turn is connected to the underfloor heating array and the hot water cylinder. The cost of the installation (excluding the underfloor heating and the hot water cylinder) was around £5,000. "We expect an average COP of 4 since the ground collector is relatively big, and average floor temperature quite low... It has a very well developed microprocessor control that has weather compensation (vital if high efficiency is to be achieved) and provides hot water up to 55°C if necessary, but [the client] seems to manage at an economy setting of about 45°C," says John Cantor. With this example, the high level of insulation means that a heating system on the first floor is unnecessary.**

USEFUL CONTACTS: Heat Pumps – John Cantor: www.heatpumps.co.uk; Ice Energy: 01865 882202; Eco Heat Pumps: 0114 296 2227; Viessmann: 01952 675000 Information – Ground Source Heat Pump Club: 0800 138 0889; Grant Information – ClearSkies: 0870 243 0930

WORDS: CINDY HARRIS AND PAT BORER

CHP is one of the few $CO_2$-cutting systems with a short payback period.

# COMBINED HEAT AND POWER

Combined heat and power produces heat and electricity from a single source, and is up to 95% efficient – but is it a worthwhile purchase for self-builders?

Combined heat and power (CHP) is essentially a means of producing both heat and electricity from a single fuel source. A bit of a two-for-one deal. You pay for the fuel to produce heat and the electricity comes free.

There are three fundamental types: internal combustion engine (the same as in your car) that run on diesel, plant oil or biogas; external combustion or Stirling engines, that run on gas (natural or LPG), oil or wood pellet; and fuel cell running hydrogen, typically extracted from natural gas, propane or methane.

There is a lot of discussion that CHP is the way of the future. A report prepared for the Energy Saving Trust in 2001 said that 13 million existing homes were suitable for CHP and that by 2010, 250,000 units would be installed each year. Unfortunately that has not happened, although the potential for cutting $CO_2$ emissions is huge.

CHP achieves around 80-95% efficiency – that is conversion of the potential energy in the fuel into useable energy – compared to less than 40% for the national grid and 70-85% for condensing boilers.

## THE DOMESTIC MARKET

Large-scale CHP has really taken off in northern Europe. District or community schemes, and even power stations, are now being built routinely as combined heat and power. But the same cannot be said of the domestic market, especially in the UK.

To the best of my knowledge, there are currently two domestic-size units readily available today: the Sun Machine supplied by Eco Solutions Ltd and the WhisperGen supplied by Whisper Tech. Many more manufacturers advertise micro-CHP but do not have a machine available for delivery today. If all they say comes true, there should shortly be lots of choice.

The Sun Machine is wood pellet fired and while is it the only carbon-neutral machine, it may, at £15,000, be just too expensive for the market.

The WhisperGen is powered by natural gas or propane and is roughly the same cost as a condensing boiler (£500 to £1,000 more expensive). If we look at a 'typical' three bedroom house, the WhisperGen is likely to return the following figures:

- Annual heat demand 20,000 kWh
- Running hours 2,700 hours
- Electricity generated 2,700 kWh
- Own use of generation up to 85%
- Unit cost of electricity 9.5p/kWh
- Value of avoided electricity purchase £218
- Unit price of electricity exported 4.6p/kWh
- Value of electricity exported £18.63
- Total value of generation £236.63
- Additional gas cost £0
- Marginal cost of unit £1,000
- Simple payback 4.2 years

Looked at in these terms, CHP is clearly worthwhile – if you have access to mains gas.

In addition, this system would reduce $CO_2$ emissions by 1.2 tons per year and the whole 2,700kWh of electricity would obtain Renewable Obligation Certificates, currently valued at around 4.5p per unit, adding a further £121 income pa to the equation. ●

### DID YOU KNOW

- **As of March 2007 it seemed that if your application for a Low Carbon Buildings Programme grant for a renewable energy installation was not in by 9am on the 1st of the month, you would miss the boat. Look for alternative sources of grant funding – Bio Energy Capital Grants, Wood Energy Business Scheme, National Parks, your local authority. Some suppliers offer the grant you would have got as a 'discount'.**
- **Recovery Insulation Ltd produces good-quality insulation material from cotton textile waste. The product has about 10% better U-values than glass fibre, is safe and soft to handle and has no toxic chemicals. The material and the manufacturing process means that it also has a low carbon footprint. www.recovery-insulation.co.uk**
- **There is concern about gas given off by chemicals used to preserve timber. Check out boron, a naturally occurring substance with a history in preventing fungal and insect attack: www.greenbuildingstore.co.uk**

WORDS: TIM PULLEN

**Natural insulation, such as Thermafleece sheep's wool from Second Nature (01768 486285), is better for your home and the environment. However, the natural option tends to be more expensive than traditional measures.**

# GOING NATURAL

## Natural insulation is better for you, your home and the environment, but is it really worth the extra cost?

Heat is money. The only way to stop that money escaping from your house is to insulate. The Energy Saving Trust tells us that the point at which the energy used to make the insulation is greater than the energy saved by the insulation is around 1m thick. The figure seems likely to vary with the material (there's very little energy in sheep's wool) but you get the idea.

Furthermore, a US Army study found that insulation gaps amounting to just 4% of the outside area of a structure equate to a 50% loss in energy. These gaps can be either literally gaps or cold bridges. If you can draw a line from the outside to the inside without cutting through a full thickness of insulation, you have a cold bridge. What these studies show us is that insulation is effective and that heat will take the line of least resistance in its struggle to leave. As to materials, we are all familiar with the usual rigid urethane foams (Celotex, Kingspan, etc) and with mineral and glass wool, which leads us to the pros and cons of natural insulation.

In general terms natural insulation achieves similar U-values to non-sustainable alternatives in a like-for-like situation, and the main types are:

**Blanket** – perhaps the most common and includes felt, sheep's wool, hemp and recycled textile waste. These are usually rolls that fit snugly between the joists and all have roughly the same U-values at 0.2 to 0.3 at 100mm thickness.

**Loose-fill** – including cork granules, wood fibre, loose fleece or cellulose pellets. These are easy to apply and especially useful for areas where joists are irregularly spaced and where pipes or other obstructions make it difficult to lay a blanket. They are also great in helping to overcome cold bridges. U-values vary.

**Rigid sheet** – includes wood fibre, straw board and cork. Can be used between rafters, underfloor, in timber fames, between joists and in cavity walls. There is very low airflow through these materials so ensure adequate air gaps. U-values are around 0.18 to 0.22 at 80mm thickness.

### THE COST OF SUSTAINABILITY

If price is your issue, then natural materials won't work for you, as they will be 20% to 100% more expensive than the non-sustainable alternative. So why use them then? Well for one thing, natural insulation does not come with a health and safety leaflet. You don't have to don goggles, breathing apparatus or gloves to work with them. And they don't gas-off. Some of the non-sustainable options are made using a variety of chemicals including ozone-depleting gases. These gases can be released over time and make their way into the property. Unsurprisingly they are not generally considered to do you any good.

Natural insulation also offers better air and moisture management – it 'breathes' more. And lastly, you can sit snug and warm in your naturally insulated home content in the knowledge that you have done as little as possible to deplete the earth's resources. ●

### DID YOU KNOW

**● You will have read in the press that the Chancellor increased grant funding by 50% in March 2007, and that the LCBP closed down, also in March, due to over-subscription, but when it re-launched in early May it was touted as "slashing household funding by 60%".**

**But hang on a minute. The LCBP closed down in response to a flood of applications for funds to install wind turbines (largely from a well-know DIY shed) that had no possibility of producing any meaningful power. This 'scale-back' will actually only apply to electricity micro-generation. Other technologies – solar thermal, heat pumps, biomass – are unaffected by the change in rules. And bear in mind that 77% of the energy used in the home is in the form of heat. Only 23% is electrical.**

**So it seems that what the Government is doing is directing funds to those technologies that generate the most energy and that will have the greatest impact on global warming.**

WORDS: TIM PULLEN

Eco-friendly paints are low in toxicity and as a result emit less odour over time – resulting in less damage both to the environment and the health of the occupier.

Clay-based plasters can use earth-based pigmentation, so there is no need for them to be painted.

# INTERNAL ECO FRIENDLY FINISHES

The way you decorate your home has a crucial effect on its eco credentials.

The internal surface finishes of a house are the parts we come into contact with most often on a daily basis. It is, therefore, important that those finishes are robust, healthy and pleasantly tactile. Eco finishes can improve the look, feel, smell and general 'ambience' of your home, and be the crowning touch to a newly built or renovated house.

## WALL FINISHES

Most internal walls and ceilings are clad with gypsum plasterboard and finished with a fine skim coat of gypsum plaster before being painted or papered. One alternative is to use a clay-based plaster applied to a clay/reed board. These self-coloured plasters (above) use natural earth-based pigments, so there's no need to paint them. They are also easy to apply as they don't 'go off' as quickly as gypsum plaster.

Lime plaster and lime wash can be used to replace cement or gypsum-based plaster, especially on old stone houses which may have used lime originally. Again, lime takes much longer than cement to set and is much more forgiving to work with. Lime washes, built up over several coats, can give a beautiful depth of colour and texture.

Timber boarding – finished with a breathable organic stain or varnish – is another healthy alternative. Fixed to battens, it can be used to cover old flaking plaster, or can be part of 'dry-lining' a wall with extra insulation. Like clay, timber is a 'hygroscopic' material, which helps to maintain a stable indoor humidity.

## FLOOR FINISHES

Fitted carpets harbour dust mites and allergens and can trigger asthma. You can still have the same degree of warmth and comfort with a natural fibre rug (wool, coir, sisal or seagrass) laid on wooden floor boards – and it's much easier to clean.

PVCu or vinyl flooring which comes as tiles or sheet material, is the cheapest smooth floor covering and is widely used in kitchens and bathrooms. But it is a major source of toxic substances and can 'off gas' a cocktail of synthetic chemicals which are very rarely tested for their effects on human health. A more attractive and durable option – though more expensive – is to use real linoleum, made with wood flour, linseed oil and pine resin on a jute backing.

## PAINTS

When we are looking at indoor pollutants which might affect our health and the wider environment, one of the main culprits is a group of chemicals known as 'Volatile Organic Compounds' or VOCs. These can be given off by the ingredients in common household paints and varnishes, and may continue to 'off gas' over an indefinite period of time.

Ideally, you should use organic paints and coatings, where the ingredients are bio-degradable, and the solvents are made from the delicious sounding 'citrus peel oil' or (not so delicious) 'gum turpentine'. Otherwise, choose paints labelled as 'low-VOC' or 'low odour'. Water-based paints have a lower VOC content than gloss or eggshell. ●

Auro Organic Paints: 01452 772020 www.auroorganic.co.uk; Completely Flooring: 0870 011 9899 www.completelyflooring.co.uk; Ecos Organic Paints: 01524 852371 www.ecospaints.com; Forbo: 01592 643777 www.forbo-linoleum.com; Green Building Store: 01484 854898 www.greenbuildingstore.co.uk; The Green Shop: 01452 770629 www.greenshop.co.uk; Construction Resources: 020 7450 2211 www.constructionresources.com; Healthy Flooring Network: 020 7481 9004 www.healthyflooring.org; Natural Building Technologies: 01844 338338

WORDS: CINDY HARRIS AND PAT BORER

**From left to right: Plaster Square energy-efficient light from John Cullen, from £104. Tel: 020 7371 9000; Eco Drop sustainable hardwood LED-based pendant lamp from Luminair, £180. Tel: 01273 383435; Havana low-energy outdoor lights by Foscarini, from Cameron Peters, £180. Tel: 01235 835000.**

# LOW-ENERGY LIGHTING

Your home's lighting scheme uses a significant amount of electricity. How do you reduce your bills without losing any design impact?

Lighting is one of the most important design elements in the home. Alongside health and safety benefits, it also sets the tone and mood, highlights features, makes the room relaxing or vibrant and as a consequence we use a lot of it. But switching to low-energy lights still seems a problem for many.

Some simple figures: a standard 100W filament bulb will cost about 50p and last about 1,000 hours. A 20W compact fluorescent lamp (CFL) will give around the same amount of light, cost about £3 and last around 8,000 hours. The CFL could save you more than £80 over its whole life and you may have 12 or 14 in your home; not to mention a saving of over one ton of $CO^2$ every year.

The big, heavy, virtually unusable energy-efficient lamps introduced some years ago have, thankfully, fallen into history, and there is now a plethora of choice in types, styles and prices. The problem then is, what kind of bulb to choose? Halogen lamps give a more intense, narrowly focused light. They are often used as downlighters and typically eight or more lamps will be needed to adequately illuminate, say, a kitchen. That is 400W of energy to achieve similar illumination to perhaps two 20W CFLs.

A 50W halogen produces as much light as a 100W filament, but a different quality of light.

Downlighters don't have to be halogen. LED lamps offer as much as 100 lumens per watt and work well as downlighters or for highlighting. They are a little more expensive but use 9W of energy for the same light output as a 50W halogen, and give a similar quality of light. Believe it or not, manufacturers often understate the lives of lamps. Those with a stated life of 8,000 hours often last over 9,000.

As with most things in life, the saying 'You get what you pay for' does tend to apply: tests by the Energy Saving Trust have concluded that brand names are actually better than cheap bulbs.

However, this does present a bit of a dilemma. Part L Building Regulations requires that 50% of the light fittings in circulation rooms (lounge, kitchen, etc) are low-energy fittings, so that we can't switch back to high-energy bulbs as soon as the inspector has left the building. These are typically four-pin, PL-type lamps which are B-rated on the Government's energy-efficiency scale. Bayonet cap or Edison screw CFLs are A-rated but fit to standard fittings and are not compliant. Part L-compliant fittings are more expensive at around £17 each compared to £2.50 for a standard fitting. ●

**Osram's Dulux Classic 10W low-energy bulbs can last as long as 12 years. www.osram.com**

**Initial Lights' Spyra-Lite GU10 energy-efficient spotlight uses only 4 watts of electricity. www.initiallights.co.uk**

WORDS: TIM PULLEN

## DID YOU KNOW

**● The Energy Saving Trust, which administers the Low Carbon Build Programme, announced on 21st December 2006 that domestic grants will now be limited to £500,000 per month. The December allocation ran out on 15th December. The situation is likely to get worse through this year with rationing of available funds to the earliest and best projects. If you currently have a planning application in, you should apply for a grant now.**

**● Proof that there is always a low-energy alternative: a 500W security light running for one hour each day will cost £22 per year to run. An 80W mercury lamp running for one hour a day will cost £3.50.**

**● Revised Part L Building Regulations, published in April 2007, require air-tightness testing for all new properties. It requires certification by an independent tester, so speak to your architect to make sure your construction method will comply.**

**Clay-based plasters can use earth-based pigmentation, so there is no need for them to be painted.**

# LIME MORTARS AND RENDERS

**Since the 19th century, cement replaced lime as the main setting agent in mortars, renders and concrete. But 'green' builders are rediscovering lime's benefits**

Cement is the most widely used material in the world, second only to water, and globally the amount used is equivalent to one ton per person per year.

## WHY USE LIME?

- Lower firing temperatures use less fuel (lower 'embodied energy') and therefore release less $CO^2$.
- Much of the $CO^2$ given off in the firing process is absorbed by lime as it sets.
- Cement kilns often burn secondary liquid fuels which may be polluting.
- Lime mortars and renders are less brittle and, therefore, less prone to cracking – generally no movement joints are required.
- Bricks and blocks set in lime mortar can easily be cleaned up for reuse.

## LIME PUTTY

Lime putty is made by adding water to quicklime, enough to make a thick creamy paste, and is the basis for all lime renders, plasters, mortars and washes. It is unwise to try and do this yourself as the procedure can be dangerous – buy it ready-made from one of the specialist suppliers. Do not be tempted to make do with the 'bagged' hydrated lime that is commonly available in builders' merchants. It is a poor substitute and will not perform well in pure lime applications. The use of pure lime products is labour intensive, needs planning well in advance and is not conducive to the quick-build, fast-track approach of the modern construction industry. On the other hand, it will be used and appreciated by those lovingly restoring old buildings in an authentic way, and by ecologically minded builders looking for a low-energy, 'soft' finish which will breathe and move with the building itself.

## HYDRAULIC LIME

Hydraulic lime is produced using certain limestones containing impurities of silica and alumina. It has the practical advantage of a quick initial chemical set (like cement) on the outer surface, followed by a much slower process of carbonation (like lime). It is, therefore, easier to integrate its use with the demands of modern construction practice.

However, this convenience comes at an environmental cost, with hydraulic lime needing firing temperatures of about 1,200°C – somewhere in between pure lime and cement – and absorbing only 50-70% of the $CO^2$ given off in the production process. Hydraulic lime has about half to two thirds of the strength of ordinary Portland cement (OPC) and uses about 30% less embodied energy.

## PREMIXED MORTARS AND RENDERS

The performance of hydraulic lime mortars and renders is dependant on using accurately gauged mixes, so many lime suppliers also supply pre-bagged lime/sand mixes.

For larger sites, under the brand name 'LIMETEC', ready-mixed dry hydraulic lime mortars can be delivered to site using tankers and bulk silos for storage. Equally welcome is the introduction of recycled aggregates, particularly the attractive 'GLASTER' using ground-coloured recycled glass. ●

USEFUL CONTACTS: Glaster – Ty-Mawr Lime Ltd: 01874 658249 www.lime.org.uk; Limetec – Lime Technology Ltd: 0845 603 1143 www.limetechnology.co.uk

**The BedZed development in Surrey, with sedum finish on the sloping roofs and grass lawns on the roof terraces.**

# CHOOSING GREEN ROOFING SYSTEMS

Roof gardens have been around for ages and until recently have been heavyweight constructions. But things have changed...

Around 25 years ago the idea of a 'turf roof' was imported from Norway, where it was common to have turfs or grass sods laid over bark shingles on pitched roofs, for protection and snow retention. The availability of synthetic rubber pond liners as reliable, low-cost roof membranes led to the grass roof becoming an almost universal symbol of an ecological building. As 'sustainability' became a more popular agenda, manufacturers of roofing products were quick to bring improved green roof products onto the market. There are now a plethora of complete systems, for drainage, water retention, membranes and plant species to cover any green roof design from flat roof gardens to steep-pitch lightweight sedum finishes.

From an environmental perspective, the green roof has much appeal:

- The natural landscape 'lost' by building a new house can be replaced on its roof.
- Ecological diversity can be increased by creating a largely wild habitat out of reach of people.
- Rainwater run-off from the roof is drastically slowed down, which is very useful in these days of increasing flash flooding.
- The vegetation growing on 30m$^2$ of turf roof will add the same amount of oxygen to the atmosphere as one person uses.
- The vegetation will have a small thermal insulating effect.

At its most basic, the green roof is just about the simplest and cheapest roof finish around. The construction will consist of a timber structure, preferably at less than 20 degrees slope (to stop the green roof finish sliding off), with a plywood deck. On this deck is placed a loose-laid single-layer membrane of butyl rubber or EPDM synthetic rubber or other materials (preferably not PVCu). For peace of mind, it is best if these membranes are professionally fitted and guaranteed, usually with a geotextile fleece, such as Terram 100, to try and retain the earth 'fines' on the roof. The simplest green finish is to lay two layers of grass turfs. Of course you may not want a grass roof – with the question of whether to mow it or not – so there are all sorts of low-lying, low-maintenance finishes available, generally called 'sedums', with many kinds of attractive wild flowers and moorland perennials. Many plant species will survive on very little soil – 50mm (2") or so – and in our experience the soil eventually washes away anyway, leaving a thick mass of roots to retain the moisture. From a building construction viewpoint, the green roof finish will protect the membrane from damaging UV light and keep it at a more stable temperature.

A more intensive roof garden will, of course, provide pleasant, high-level extra outdoor space – very useful in areas of dense urban housing. This sort of green roof will generally have a thicker soil layer and be laid nearly flat. Many products are available which help to retain more water – usually an 'egg box' layer which holds little pockets of water. ●

USEFUL CONTACTS: Prelasti Roof Garden (Pirelli EPDM rubber): www.prelasti.com; Alumasc 'ZinCo' Green Roof System: 01744 648400 www.alumasc-exteriors.co.uk; Phoenix 'Resistit' EPDM membranes: www.phoenix-ag.com; Alwitra 'Evolastic' PVCu-free membrane and 'Eva-Gro' system: www.alwitra.de

WORDS: CINDY HARRIS AND PAT BORER

# TURFING YOUR ROOF

**While soil is not a good insulator it can provide thermal mass to stabilise temperatures, and you don't need to mow.**

Green or 'living' roofs are roofs that are intentionally vegetated with a planted surface – primarily grass or turf – in a soil depth of around 50-100mm. Grass roofs are the simplest option and do not even require mowing — unless you hanker after a bowling green finish! Untended turf roofs will have a rough, slightly shaggy appearance; they will die back in dry periods but re-grow later in the year. Wildflowers and bulbs can be planted in the soil and local species will often colonise. They will need just occasional weeding. More recently, cultivated species such as sedums have been used, which are drought tolerant and shallow rooted, and so ideal for these exposed conditions.

Roofs are usually low-pitched (a maximum of 15 degrees from the horizontal) to stop the soil from sliding off until the roots are established.

Pros and Cons

Many claims of eco-friendliness are made for green roofs, but not all of them are justified. Soil – especially when wet – is not a good insulator, though it can provide useful thermal mass which helps to stabilise temperatures. However, you will still need as much insulation as possible between or under the rafters. In spite of their 'low-tech' appearance, all green roofs depend for their performance on a high-energy, high-impact plastic or rubber membrane, which adds significantly to the roof's overall environmental impact.

**Left: This planted sedum roof in Oxfordshire has two high-performance membranes beneath it that cover 120mm of insulation placed above a vapour barrier.**

However, there are micro-climatic benefits, particularly in cities, from any area of green growth: vegetation will absorb $CO^2$ and release oxygen, aid bio-diversity, improve air quality by absorbing pollution, and reduce the 'heat island' effect by evaporation in hot weather.

In rural areas, green roofs can help mitigate the impact of development by merging a new building into its surroundings. And in any area which is prone to flooding or with an inadequate drainage system, green roofs will help avoid problems by significantly delaying or preventing rainwater run-off. ●

### HOW DO THEY WORK?

**The turf or sedum mats rest on a waterproof membrane which is laid loose over boards fixed to rafters. A protective layer may be used to stop the membrane being punctured.**

USEFUL CONTACTS: Glaster – Ty-Mawr Lime Ltd: 01874 658249 www.lime.org.uk; Limetec – Lime Technology Ltd: 0845 603 1143 www.limetechnology.co.uk

### DID YOU KNOW

**● Despite the Government's U-turn on making Home Condition Reports an essential part of the new Home Information Pack – which vendors of houses of certain sizes had to provide from 1st June 2007 – the Energy Performance Certificate, which gives potential buyers information on a home's energy efficiency, will still be a critical part of the pack. "It will help everyone respond to the global challenge of climate change," says Housing Minister, Yvette Cooper.**

**● As part of a wider review of Permitted Development Rights, the Government will soon be removing the need for homeowners to apply for planning permission to install energy-generating features such as wind turbines and solar panels.**

**● Planning Policies for Sustainable Building is the name of a new report giving guidance to help councils adopt policies to ensure the building of low-impact new homes.**

WORDS: CINDY HARRIS AND PAT BORER

SIPs roofs have in-built insulation, no rafters, and can cross large spans.

Renewable insulation materials: flax, sheep's wool and cellulose.

# INSULATING YOUR ROOF

**With new legal requirements stipulating a minimum thickness of insulation it's important to look at ways to insulate new and existing roofs.**

Whether renovating a roof, or building a new one, there is now a legal requirement to insulate to the levels of Part L1 of the Building Regulations. Many 'green builders' consider these levels a minimum but, nonetheless, the thickness of insulation now required is around 270mm for a conventional loft, and 200mm for 'room in a roof'. Ecological homebuilders should aim for 300mm–450mm.

## THE LOFT

While it is easy to insulate a conventional 'cold' loft, it is not a satisfactory form of construction. The thick layers of insulation prevent the loft from fulfiling its traditional role as storage space unless a separate floor is fitted. Having water pipes and tanks in the loft space can be a disaster, bringing possible heat loss and condensation problems. The loft hatch, which should be insulated to the same level as the rest of the loft, is an opening that is difficult to seal.

Any fibrous insulation material can be used, with one layer between the ceiling joists and a second at 90° over the top. All fibrous insulants have the same thermal value, so it is the thickness of material, or how much you put in, which determines how well the roof will be insulated. The choice of material comes down to relative environmental and personal 'friendliness', set against cost. Flax (NatalinFlax) and sheep's wool (Thermafleece) are the 'green' alternatives to mineral wool (Crown Glass, Rockwool), but the latter are the only ones available with the HEAT Project grant (www.heatproject.co.uk) as they are significantly cheaper. Another popular eco option is cellulose insulation (Warmcel). This is made from recycled newsprint treated with flame retardant and a mild biocide. It can be bought in pre-fluffed form, in bags, and simply spread out where it fills every gap and hollow. For larger jobs you need a professional installer and machine to blow it in.

## ROOM IN ROOF

Many self-builders are choosing to build rooms in the roof. The advantages are: the maximum use of the house volume; a more airtight construction, with all services within the heated envelope; and interesting room shapes with sloping ceilings. The best constructions do not use solid timber rafters; at over 200mm deep these would be uneconomic, and form undesirable thermal (cold) bridges. Timber I-beams (eg Masonite, Trus Joist) or the DIY version made from spaced rafters are the preferred construction that can be used with any type of fibrous insulation. Alternatively, conventional rafters, filled with insulation, can have insulation boards fixed either over or under them to provide the necessary levels of insulation. The board materials can be from renewable sources (e.g. Pavatex Diffutherm), or high-performance, zero ozone-depleting foams that can be used to form a more slender construction (e.g. Kingspan Thermapitch). Some ecological homebuilders prefer to use a material from renewable sources, but there is an argument that it is better to use oil and coal to manufacture high-performance, long-lasting insulants, rather than burn it up as fuel.

An alternative way to construct an energy efficient roof is with Structural Insulated Panels (SIPs). SIPs have no rafters; a high-performance foam is bonded to two layers of orientated strand board to make a strong panel that can span from eaves to ridge. ●

USEFUL CONTACTS: Grants for insulation – HEAT Project: 0800 093 4050 www.heatproject.co.uk; Warm Front: www.eaga.co.uk; Timber I-beams – Masonite – Panel Agency Ltd: 01474 872578; Trus Joist: www.trusjoist.com; Warmcel – Excel Industries Ltd: 01495 350655 www.excelfibre.com; Thermafleece – Second Nature UK Ltd: 01768 486285; from the Green Building Store: 01484 854898; NatalinFlax – from Ty Mawr Lime: 01874 658249; Pavatex Diffutherm – from Natural Building Technologies: 01491 638911; TRADIS – Timber Frame Solutions Ltd: 01670 798700; www.timberframed.com; Excel Industries Ltd: 01495 350655; Mineral fibres – Rockwool Ltd: www.rockwool.co.uk; Crown Glass Mineral Wool: www.knaufinsulation.co.uk; SIPs roofs – Kingspan Insulation: www.insulation.kingspan.com; Milbank: 01787 467299

WORDS: CINDY HARRIS AND PAT BORER

# STRAW BUILDINGS

**Straw bale buildings are a rarity in the UK, but are a sustainable, easy-to-build and well-insulated construction form.**

In these days of increasingly technological solutions to the question of how to build, it is unusual – and refreshing – to find a simple, plant-based building material that can provide the structure and/or the insulating infill walls for several types of low-rise buildings.

Straw is a renewable raw material that is a by-product of grain production, and in many arable areas is a waste product that is not easy to dispose of — meaning straw bales are low-cost and easily available.

## HOW DOES IT WORK?

Walls made with straw bales can be fully load-bearing or the bales can be used as an infill between timber frames. In either case the bales should be as tightly compacted as possible.

A strip foundation and plinth wall up to DPC level is laid in the normal way but 450mm wide, and bales are then stacked like giant building blocks, with staggered joints. Each course is pinned to the one below with timber or metal rods. Door and window openings are sized as multiples of bale lengths (900mm) or half lengths and lined with framed plywood. Timber plugs or wedges for wall fixings and sockets can be hammered in as the walls go up. The bales are then rendered inside and out, ideally with a lime render or clay plaster. Roof construction is conventional, with the rafters or trusses supported on an extra-wide wall plate.

This is an ideal method for self-builders, but beware the ease and speed of construction – it can lead to a condition called 'bale frenzy' where people get so carried away with their success that they forget to check for plumb and straightness!

## PROS AND CONS

One of the main advantages of straw bale construction is that you will get a highly insulated wall simply by virtue of the bale widths and, therefore, the wall thickness. Typical U-values quoted are from 0.21-0.13 $W/m^2K$. Straw bale walls also provide relatively good sound insulation and, contrary to popular belief, are very good at resisting the spread of fire. It is in fact surprisingly difficult to set fire to a well-compacted bale, and the render or plaster coat will provide additional protection. Small rodents can be kept out of what might be a tempting nesting site by keeping the bales well clear of the ground with physical barriers installed if necessary.

The only real problem in building with straw – but it can be a big one – is keeping the bales dry at all times. Walls under construction must be well protected from moisture – in a framed building the roof should go on first and the walls be built up underneath it. If any bales do get wet they should be replaced immediately or they will start to compost.

Good design should help to keep the bales dry throughout the life of the building – by providing overhanging eaves and a generous distance between ground level and DPC (150mm is the minimum, but 200-300mm is better), to avoid water penetration and splashback.

If you are sceptical about the performance or durability of straw bale, you could try using it first for an ancillary building, such as a garage, studio, workshop, or even stables. ●

**Top: Eco self-build house in West Wales with straw bale walls which are partly load-bearing, partly infill around an oak frame.**
**Left: Straw bales being stacked around a load-bearing timber frame.**

USEFUL CONTACTS: The Straw Bale Building Association: 01442 825421 www.strawbalebuildingassociation.org.uk

WORDS: CINDY HARRIS AND PAT BORER

This new home was 'stick-built' using a hybrid version of a closed-panel breathing wall design, 300mm thick, made using Masonite engineered timber I-beams, clad either side in OSB and filled with cellulose insulation. It has an SAP rating of 97.

This timber frame self-build uses spaced studs and rafters — note the prepared OSB 'nailplates'.

# CHOOSE TIMBER FRAMES

## The humble timber frame is one of the most energy efficient ways to build.

Wood is very much the material of choice for eco-friendly self-builders. It is renewable, carbon efficient and used properly can result in an airtight structure that can be filled with insulation. Timber stud framing can also be constructed entirely on site – 'stick-built'. The studs and wall plates are cut, panels are assembled flat on the ground – sheathed to keep them square – and raised up. A floor 'platform' of joists is fitted and the first floor and roof constructed off this platform. Windows are fitted and the construction is then insulated and finished.

Today, there is a trend towards factory-made 'cassette' timber framing. The wall, floor and roof panels are constructed in factory conditions. They are often insulated, have the windows and doors fitted – and can even be pre-wired and plasterboarded. The whole 'kit' arrives on site and can be erected and made weatherproof within a day. Externally, the timber frame is usually finished in a brick or block veneer tied back to the frame, or a render finish can be applied directly onto laths. Apart from the advantage of fast erection times, factory-made systems should be better constructed and more airtight than 'stick-built' structures.

Until recently, the timber stud panels were made of 100mm x 38mm studs filled with mineral wool, sheathed on the outside with 9mm plywood and finished internally with a vapour barrier, plasterboard and skim plaster. This construction no longer meets the Building Regulations' thermal standards (Part L), and there is a move towards fitting the sheathing on the inside to create the inherently safer 'breathing wall' construction. To avoid air leakage and other problems, electrical services are best placed in a void formed by fitting 25mm battens to the inner sheathing. There are many different constructions available that comply with, or preferably exceed, the current Part L requirement (U-value of 0.35W/m$^2$K):

● **Deeper studs.** A wall using 150mm x 38mm studs will comply with Part L, after allowance is made for the 'thermal bridging' of the timber studs and plates.

● **External sheathing of studs with rigid insulation.** This method has the advantage of being an easy improvement that still uses conventional (and cheap) studwork (eg the APS and NBT 'NewBuild' systems).

● **I-beam studwork.** The poor thermal properties of a solid timber stud are overcome by having thin webs joining two 50mm x 50mm studs (eg the Masonite Beam and TRADIS Systems, Finnjoists and Trus Joists).

● **Structural Insulated Panel systems (SIPs).** A sandwich of rigid foam insulation glued between two sheathing boards makes a rigid panel for walls and roofs. As there are no studs, and as the foam is a high-performance insulator, SIPs wall panels (eg Kingspan TEK) can be quite slender compared to their conventional equivalent.

● **Double (spaced) studs.** For very high levels of insulation (eg 300mm – U-value 0.13W/m$^2$K), a good solution is to construct two conventional stud walls, spaced apart the required distance by plywood nail plates. ●

USEFUL CONTACTS: APS – Advanced Panel System by Selleck Nicholls: 01579 370740; NBT NewBuild – Natural Building Technologies: 01491 638911; Masonite Beams – Panel Agency: 01474 872578; TRADIS: Excel Industries Ltd: 01685 84520; Finnjoists – Finnforest: 0800 004444; Trus Joist: 01214 456666; TEK Building System – Kingspan Insulation: 0870 850 8555; Urbane Sustainable Homes: 0117 955 7224; Centre for Alternative Technology: www.cat.org.uk

WORDS: CINDY HARRIS AND PAT BORER

# THE END OF OPEN WINDOWS

**The new airtightness regulations demand not only mechanical ventilation systems, but also a complete lifestyle change for energy savings to work.**

It is mandatory for all houses gaining planning consent after April 2006 to meet the Building Regulations Part L150 standard on airtightness. That standard is a 'leakage' of not more than 10m3/h/m$^2$@50pa. That is, 10 cubic metres of air per hour escaping for every square metre of the envelope surface area with the air at a pressure of 50 pascals. What that actually means is a bit of a mystery, and whether it is a good idea is another thing altogether. We are being forced to adopt airtightness standards, but not being told that without mechanical ventilation, dehumidification and effectively air conditioning, we will be living in a warm, damp, foul-smelling atmosphere polluted with the chemicals gassed off from construction materials.

The 2005 Building Regulations stated that we had to have a maximum of 1.5 air changes per hour (ach). It was in the late 1980s that the standard changed from a minimum to a maximum. Until 2005 1.5ach was considered to be the minimum amount of air needed to maintain oxygen levels, and clear out exhalation $CO_2$, pollutants and smells. The 2006 regulations are based, at least in part, on the Canadian R2000 standard which calls for 0.8ach at 50pa.

When asked why 10m$^3$ was chosen as the standard, Mr Colin King, head of BRE Wales, who was influential in drafting the standard, said it is "an arbitrary figure. It has no direct relationship to air changes per hour." But of course there has to be a relationship and the 10m$^3$ standard is said to be around 1.2ach. And to put that in perspective, the German PassivHaus standard calls for 'less than 1ach' but again at 50pa.

The change comes as a result of a Government fact-finding mission to Canada and Sweden in 2002 to figure out how to make our houses more energy efficient. They came back with the clear idea that controllability was the answer. What we had was uncontrolled air movement through trickle vents and air bricks, and what Canada and northern Europe had was air movement controlled by mechanical ventilation. The model for our new standard was Canada's R2000 E-House, which also mandates that mechanical ventilation is a requirement in all homes.

At 10m$^3$/h (or 1.2ach) it is possible that the building will be 25% more energy efficient than the 2005 standard. If we achieve the target standard of 5m3/h, BRE say that we will save 40% of our heating energy consumption. To make the standard work we need two things: a house that remains sealed, and a reliable mechanical ventilation system. To not put too fine a point on it, with permeability at even 5m$^3$/h, if the ventilation system breaks down there is a good chance of suffocating – which is why the Canadian standard mandates for ventilation systems and that they must be connected to mains electricity.

The Canadian standard is based, quite reasonably, on the requirements of the Canadian climate. Canadian temperatures lead to a dry climate (moisture gets frozen out of the air in the long cold winters). So much so that the mechanical ventilation systems inject moisture into the air to prevent excessive drying in the house. This is not the British experience. In fact, our problem will be the reverse. To quote Colin King again: "It is a lifestyle change. People will have to learn to not open their windows and not dry clothes indoors."

What happens to those potential energy savings if we choose not to change our lifestyle? If we live in the hermetically sealed houses now demanded in the same way as we do in our current draughty properties, the energy savings literally go out the window. We have a habit of enjoying opening windows and moving freely in and out of the house, without having to pass through a double-door air-lock. That habit has developed because we live in a temperate, relatively damp climate. I for one would not be happy if I could not open the windows and listen to birdsong while I sit at my desk. ●

WORDS: TIM PULLEN

New walls need to work in much the same way as old walls, which allow fresh air to enter the house and stale air to exit, reducing the possibilities of condensation.

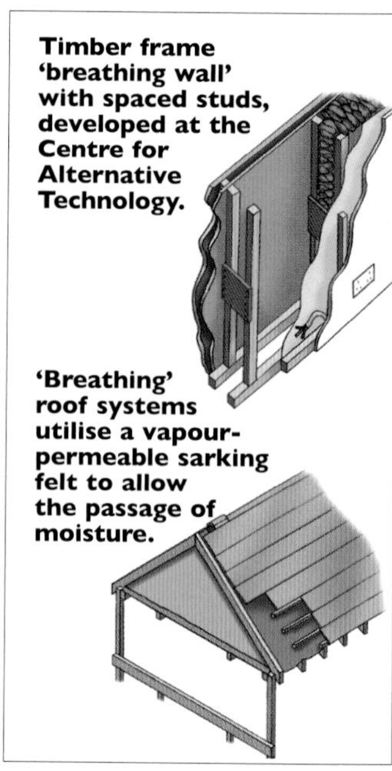

Timber frame 'breathing wall' with spaced studs, developed at the Centre for Alternative Technology.

'Breathing' roof systems utilise a vapour-permeable sarking felt to allow the passage of moisture.

# AVOIDING CONDENSATION

**The only way to reduce the possibility of condensation is to allow walls and roofs to 'breathe'.**

We have all seen condensation running down windows or walls, especially in poorly insulated kitchens and bathrooms. In well-insulated houses with double glazing, this surface condensation is less likely to occur, as the surfaces will be relatively warm. Instead, moist air will try to migrate slowly through the walls, and the danger here is that it may condense into liquid if it hits a relatively cold surface within the wall construction. This is known as interstitial condensation, which can cause timber studs or rafters to rot.

The conventional solution to avoiding interstitial condensation within timber walls is to install a 'vapour barrier', or polythene sheet, on the inside of the insulation, behind the plasterboard. This will work provided the polythene is kept intact. In practice, however, holes are made to get services through, or it is not properly sealed around openings, meaning moisture can then get into the wall cavity – a recipe for disaster!

Other, more reliable, solutions have recently become available which accept that some moisture will get into the wall, but which make it very easy for any moisture inside to migrate to the outside and evaporate. In this approach, a 'vapour control layer' (VCL) is used in place of the vapour barrier. This VCL is roughly five times more vapour resistant than the outer sheathing and could be ply, hardboard, foil-backed plasterboard, or a suitable building paper. The outer layer needs to be porous and 'breathable', such as bitumen impregnated fibreboard.

The outer sheathing can also provide some racking resistance for the timber frame and may eliminate the need for diagonal bracing. A ventilated space (min 25mm) should be left between the sheathing and the wall finish. A 25mm 'services zone' behind the plasterboard will help to ensure that the VCL stays intact.

Ideally, the insulation material should be 'hygroscopic', or able to regulate humidity levels by absorbing excess moisture. Examples are cellulose fibre, such as Warmcel 500, or sheep's wool, such as 'Thermafleece'.

Breathing roof constructions are important too. Most traditionally constructed 'cold' roofs with ventilated loft spaces are in no danger of harbouring condensation. There are now several vapour-permeable sarking felts available (eg Klober 'Permo') and these should always be used on new or replacement roofs.

In spite of the title, the wall or roof does not actually 'breathe', as it is moisture rather than air which is passing through it. It is, however, similar to us wearing a 'Goretex'-type coat, instead of an impermeable plastic coat which makes us sweaty because it will not allow moisture through. ●

USEFUL CONTACTS: Cellulose fibre insulation – www.excelfibre.com/downloads/products/Warmcel-brochure.pdf; www.excelfibre.com/downloads/concepts/Sustainable-social-housing%20.pdf; Sheep's wool insulation – Second Nature: www.secondnatureuk.com; Klober – www.klober.co.uk; Breathing construction – Natural Building Technologies supplies the components for a breathing construction system: 01844 338338.

# LOW ENERGY CAVITY WALLS

## Which form of wall construction is best in terms of its ecological credentials?

For the last 75 years, the external masonry cavity wall – two skins of brick or block with a space in between – has dominated domestic construction. The cavity prevents damp penetration and provides thermal insulation. Nowadays, cavities are partially or wholly filled with thermal insulation to meet, or exceed, the current building regulations (100mm of insulation until 2008's revision). The regulations allow cavities up to 300mm (12") wide using conventional wall ties and many low-energy houses now have insulation up to this thickness.

Masonry materials such as brick, stone and concrete have high 'thermal mass', which means that heat can be absorbed, stored and released depending on the ambient temperature, providing a fairly constant internal temperature. It is now thought that providing thermal mass within a house will be increasingly important as the climate changes to, perhaps, warmer summers.

The modern practice of dry-lining the inside face, by fixing plasterboard sheets to it, is to be discouraged: it is difficult to seal against draughts, and it prevents the internal masonry leaf from having any useful function as primary thermal mass. Instead, traditional plaster or render should be applied straight on to the wall. Floor joists should be supported by steel hangers and separate lintels provided for each leaf of the cavity, to prevent cold bridging. Cavities should be closed at openings by timber or rigid insulation fill, the inner leaf should not be 'returned' in masonry as this would be another cold bridge.

### MASONRY MATERIALS

Walling of earth from the site – cob, or stabilised earth blocks – is the best ecological choice for a masonry wall, for embodied energy and health reasons. Fired clay bricks, however, make an attractive, durable and low maintenance outer skin, but they must be laid in a lime-rich mortar, so they can be reused in the future and also to minimize any cracking problems. Second-hand bricks are the best choice. If not, new bricks should be sourced as locally as possible. It is difficult to avoid the ubiquitous concrete block; it is best to choose those now made using recycled and secondary aggregates, or those using cement replacements such as PFA.

In the rest of Europe, the more environmentally friendly hollow clay blocks are universal. They are now being imported to the UK and some are now being manufactured here by Ibstock. Another low-impact material is sand-lime brick – autoclaved at lower temperatures than clay brick, and hence having a lower embodied energy. Most Dutch houses are constructed using pre-cut sand-lime panel blocks (left), a fast and environmentally sensible practice that has yet to reach our shores.

**The Building Regulations now allow cavities of up to 300mm wide using conventional wall ties. BedZed housing in Surrey is 'zero-$CO_2$' with 250mm rock wool-filled cavities.**

### INSULATION MATERIALS

Cavity walls can be insulated using a fully filled loose material, such as blown glass wool or rock wool, or poured perlite beads. Filling is done after the walls are constructed to avoid constructional and energy conservation problems associated with trying to insulate a wall as it is being built. Rigid insulation boards made of plastic foams, such as urethane and phenolic foam offer thermal benefits (less thickness for the same insulating value) and now have zero ozone depletion potential. However, they must be installed properly without gaps to ensure airtightness. ●

USEFUL CONTACTS: Hydraulic lime – St. Astier, from e.g. Ty Mawr Lime: 01874 658249; Sand-lime bricks and CVK panels – Esk Building Products: 01228 527621; Extruded cob earth blocks – Mike Wye: 01409 281644; Perlite insulation – Silvaperl Lightweight Aggregates: 01427 610160

# RAMMED EARTH WALLS

**Rammed earth is a traditional form of construction that is regaining popularity today thanks to its sustainable credentials.**

Rammed earth, or 'pisé de terre', is an ancient and worldwide form of wall construction that is blossoming today. In China, Yemen, Nepal, Egypt and Morocco the technique has been used for millennia; the Roman armies introduced rammed earth building to the south-east of France. In central Europe, earth building flourished in the 19th century and was revived after the first and second World Wars – in the UK by Portmeirion architect Clough Williams-Ellis.

The tallest earth building in Europe is in Weilburg, Germany. Built in 1828, it has seven storeys and rammed earth walls tapering from 750mm to 400mm thick. A whole village – Domaine de Terre – was built of earth in the 1980s. In Australia there is today a thriving rammed earth construction industry building hotels and houses, whilst in Europe there is a small band of rammed earth experts creating beautiful earth structures.

So why are an increasing number of self-builders with eco principles in mind interested in using it? Not only is it sustainable thanks to its obvious natural occurrences, but new research is also beginning to emphasise its qualities as a solid form of construction: think of rammed earth as 'instant rock'. The natural processes that make sedimentary rock over thousands of years are imitated by the earth rammer. Rammed earth construction consists of moist, loose subsoil highly compacted between shuttering, in layers of 100-150mm depth. The mechanical compaction compresses this depth to about half, and forces the clay molecules to bond with the various aggregates. The shuttering is then moved along or upwards to form a whole wall. Because of the relatively dry mix, the shrinkage of rammed earth elements is much less than for other earth-building methods, and the strength is correspondingly higher. A rammed earth wall will dry out and become as tough and beautiful as sandstone, as long as it is protected from damp.

Knowing the exact composition of the soil and the correct amount of water to be added is critical for the success of this method. Usually the builder will take the material found on site and adjust its composition by mixing in other 'bought in' materials. Where there is no suitable material on site, a mix can be designed and assembled from available quarry wastes and clays.

**Rammed earth walls and chimney by Martin Raunch, Austria.**

At the Centre for Alternative Technology, tests were carried out on various samples of local quarrying 'overburden' to determine the composition and strength of the material. The best samples showed a compressive strength of 2.29 Newtons/mm$^2$ – well in excess of what is needed to support a lightweight, two-storey building.

Where stabilisers such as lime or cement are considered necessary, these should be kept to a minimum. Where they are used routinely, as with most modern earth construction in Australia, some of the environmental benefits are reduced.

Earth mixes can be compressed in a block-making machine. Blocks are produced in standard sizes and allowed to dry, under cover, for several weeks. They can then be laid in a lime- or clay-based mortar and rendered with the same. Stabilised earth blocks are made stronger and more durable by the addition of small amounts of lime or cement (5-10%).

Whilst rammed earth construction has not yet reached the commercial scale found in Australia, much useful research has been carried out in universities, and there is now a DTI handbook, signifying its growing popularity as a sustainable form of housebuilding. ●

USEFUL CONTACTS: Rammed Earth Design and Construction Guidelines, Peter Walker et al. – BRE Bookshop: www.brebookshop.com; Rammed earth designer/contractors – Roland Keable/In Situ Rammed Earth Co. Ltd: 020 7241 4684; Simmonds/Mills: 01952 433252

### COBTUN HOUSE, WORCESTERSHIRE

In this recent RIBA award-winning house, an encircling earth wall, which appears to crumble or decay at its southern end, encloses the site. The cob wall construction uses site soil and locally sourced clay, with straw acting as the binding ingredient. Following the wall round, an entrance leads through the massive wall to the front door and circular top-lit hall. The wall continues past the entrance, the house abutting its south face in horizontally boarded oak and glass.

Earth for cob walling, excavated on site, was predominately fine sand with layers of gravel and clay. A study by the Centre for Earthen Architecture, University of Plymouth School of Architecture, recommended adding an additional 25% of imported spoil from another of the contractor's nearby sites. As the majority of earth is from the site itself, and as there are virtually no transportation or manufacturing costs, the 'embodied energy' of the material is almost zero. Architect: Associated Architects

# COB AND LIGHT EARTH WALLS

**Which form of wall construction is best in terms of its ecological credentials?**

Earth is one of the most immediate and locally available materials it is possible to build with. It is also one of the cheapest and lowest environmental impact construction methods, and certainly the one with the longest history. Over one-third of the world's population live in houses built from earth. In central Europe, earth building flourished in the 19th century and was revived after World War I and World War II – in the UK by pioneering architect Clough Williams Ellis. But why should it appeal to today's 21st century self-builder?

In general, earth construction involves a very low external energy input, and creates virtually no pollution. Earth buildings have a high 'thermal mass', meaning they can absorb and store solar energy, and re-release it in the form of heat when the building cools down. Earth is not particularly good at insulating, but its performance can be improved by adding organic fibres or lightweight mineral aggregates. On the whole though, it is not suitable for external walls without additional insulation. Being hygroscopic, earth walls can also regulate humidity levels, which is thought to be an important factor in achieving good indoor air quality.

Most construction activity begins with clearing earth away from the site of the building – it is this excavated material that can be used in the construction, although it is the subsoil rather than the topsoil that is required.

## COB

This is the main form of earth building to be found in Britain, in parts of East Anglia, the East Midlands and West Wales ('clom') – but particularly in Devon, where hundreds of Devonshire cob houses are still standing, the earliest dating back to the 15th century. In this method, the sub-soil is mixed with straw and water, and then pounded or trodden until it reaches a suitable consistency. It is then laid in horizontal layers, and again pounded down, to form free-standing mass walls.

## LIGHT EARTH

In this country, cob is a form of light earth construction, though using more earth than straw, while the original 'Leichtlehm' or 'Light Earth' was developed in Germany in the 1920s using straw in a clay slurry. Recent developments have incorporated lightweight mineral aggregates or plant fibres, in an attempt to increase the thermal performance of earth walls to meet stricter building regulations – even then, earth walls have to be very thick (about 800mm) to meet current required U-values. These lightweight mixes are usually used as infill in timber frame construction. ●

**The first UK building made from light earth walls using a mix of blocks, monolithic clay and straw within a timber frame.**

# MAKING OLD WALLS WARM

The walls of your house should be as well insulated as possible, so make sure you know all the options.

The external walls of a house are where most heat is lost, but there are a range of solutions for old houses regardless of the construction type.

Most of the external 'envelope' of a house consists of walls and this is the area through which most heat is lost. New houses will have been built to reasonable standards, but older, un-renovated properties will be poorly insulated, if at all. The method of insulation, and the materials used, will depend on how the wall is constructed. Regardless of your home's situation, there are a variety of options for reducing heat loss.

## CAVITY MASONRY WALLS

In general, houses built in the last 50-60 years with brick or concrete block walls will have a cavity (usually 50mm deep) between the inner and outer 'skins' of brick or block. You can check whether your house has cavity walls by measuring their thickness at door or window openings – brick cavity walls are at least 265mm (10") thick. Where these cavities have not already been filled with insulation (most likely in older houses), cavity insulation is a very effective way of reducing heat loss. A water-repellent insulation material, such as granulated Rockwool or polystyrene beads, can be injected into existing cavities by professional installers. Where there is a door or window opening, the insulation should extend up to the frame, to avoid 'cold bridging'.

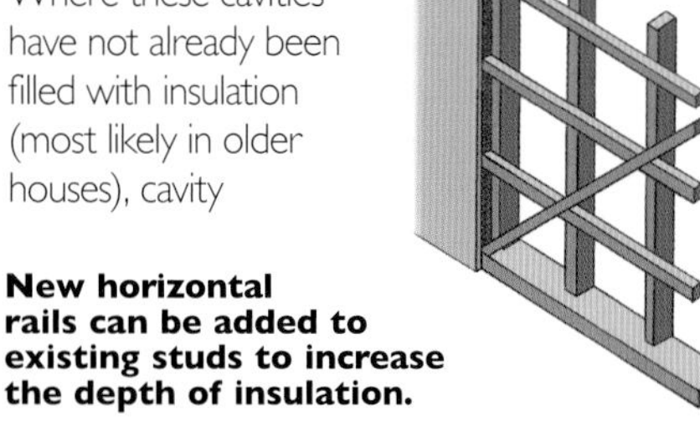

**New horizontal rails can be added to existing studs to increase the depth of insulation.**

## SOLID MASONRY WALLS

These are likely to date from before the mid-20th century and be made of stone, brick or earth, with thicknesses from 600mm upwards for stone or earth. Solid brick walls are usually two bricks deep or 225mm (9"). They can be recognised by the appearance of a half brick – or full brick end-on – in the bond of the brickwork.

This type of construction can be insulated on either the outside or the inside. The former is more expensive but most effective, consisting of render applied to an insulating board that is fixed to the outer wall, and is generally carried out by professionals. Internal insulation, or 'dry-lining', can be done easily on a DIY basis and consists of constructing a timber frame (minimum 100mm deep) inside the external wall, filling it with a soft insulation quilt, followed by a vapour-control layer, plasterboard and finish. A major disadvantage of this method is that quite a lot of the internal room area is lost. A less effective solution (because there is less depth of insulation) is to fix new 'thermal' plasterboard with a bonded layer of foamed plastic, directly to the wall.

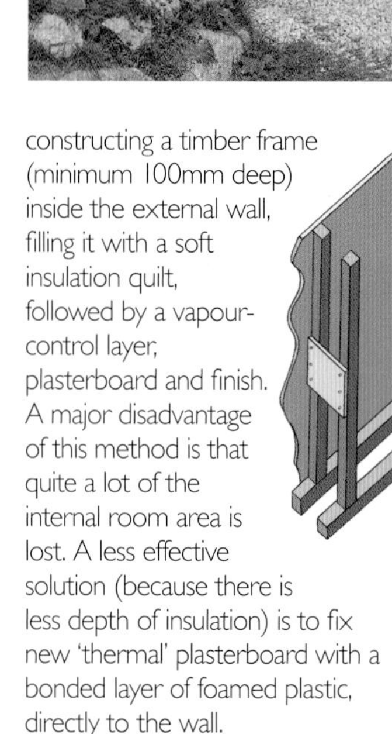

**'Spaced stud' timber frame, using plywood plates to separate old and new studs.**

## TIMBER STUD WALLS

Because timber walls are kept dry, they can take more benign forms of insulation such as sheep's wool: 'Thermafleece', or recycled cellulose fibres: 'Warmcel'. The latter can be injected into existing cavities through holes drilled in the plasterboard. Alternatively, the internal or external covering to the studwork can be stripped off and sheep's wool or mineral fibre quilt laid upright between the studs. In the process, it is worth considering increasing the depth of the insulation by adding extra timbers, fixed either horizontally at right angles to the studs, or vertically and spaced off from existing studs. A less disruptive way of increasing insulation would be to follow the dry-lining method described earlier. This applies equally to timber walls which have an external brick or block outer skin.

When installing or improving insulation in a timber framed wall, it is important to prevent condensation occurring within it. ●

USEFUL CONTACTS: Energy Saving Trust: www.est.org.uk/myhome/insulation; Cavity Insulation Guarantee Agency: www.ciga.co.uk; National Insulation Association: www.nationalinsulationassociation.org.uk; Cellulose Insulation: www.excelfibre.com Sheep's wool insulation: www.secondnatureuk.com/info.htm

WORDS: CINDY HARRIS AND PAT BORER

**Natural Solutions designs and builds dual-chamber compost toilets, and supplies urine separators and DIY components.**

**AVERAGE WATER CONSUMPTION IN ONE YEAR**

| | | |
|---|---|---|
| Flushing the toilet | 33% | £132 |
| Washing machines | 21% | £84 |
| Baths and showers | 17% | £68 |
| Kitchen sink | 16% | £64 |
| Washbasin | 9% | £36 |
| Dishwashers | 1% | £4 |
| Hosepipes | 3% | £12 |

# REDUCE YOUR WATER WASTAGE

**Flushing the toilet costs the average household £132 a year – but that cost and other water wastage could be halved with a few eco-conscious purchases.**

How can it be, in a country where we complain about the rain all the time, that the South-East was suffering a hosepipe ban so early in 2007? Yes, it had been an exceptionally dry year, but the real culprit was domestic water use, which has doubled from 75 litres/person/day in 1960 to 150 litres/person/day – and water suppliers have not been able to keep up with the demand.

As a resource, water is being used at an unsustainable rate – underground aquifers are being emptied faster than they are replenished. Quite understandably, there is a significant charge for your water supply and for treating the sewerage water leaving your house, which has risen steeply in the last decade to around £400 per year for a typical house.

Although most people dislike having one fitted, water meters allow you to check how you are succeeding on the water-saving front, usually causing water consumption to drop considerably, by around 30%.

There are a number of ways to reduce our overall water consumption, and to use water more efficiently. New toilet cisterns are required to be no more than six litres, but most existing ones are nine litres. As most siphon toilets will flush perfectly well with less than nine litres of water, a water-filled plastic bottle in the cistern will make significant inroads into what is probably your largest water use. A 4.5-litre siphon flush toilet, approved by the Water Regulations Advisory Scheme (WRAS), is now available in the UK. Such a toilet will halve water used for flushing, and therefore could save around £70 per year. There are also dry compost toilets, which avoid water use altogether. Compost toilets not only use no (expensively processed) water, but they also produce a useful fertiliser.

A bath uses about 110 litres of water, a shower 30-40 litres. As they are so quick and convenient we tend to take more showers than baths, which evens out the score somewhat. A power shower, however, can use as much water as a bath. Further water reductions can be achieved by installing low-flow shower heads which aerate the spray more effectively, so you get just as wet with less water.

As for dishwashing, it is debatable whether machine- or hand-washing is more water-efficient. Most dishwashers use less than 20 litres on a full load. A standard washing-up bowl holds seven litres and a standard sink, around 15 litres. The way in which you wash up will be crucial – for example, if you rinse the dishes in another sink full of water, you will be using more water than a dishwasher. According to the Environment Agency, the major environmental impacts of dishwashers are related to detergents rather than water or energy use. Bear in mind that 'Economy' or 'Half Load' settings typically reduce water and energy use by only 10-25%, and, as with all appliances, purchase one with an 'A' rating. ●

USEFUL CONTACTS: Compost Toilets – Natural Solutions: 01686 412653 www.natsol.co.uk; Low-flush Toilets – Kingsley Clivus: 01837 83154; The Green Building Store: www.greenbuildingstore.co.uk

# COLLECTING RAINWATER

Rainwater harvesting could ease both your conscience and your water bills, but what are your best options?

It is always better to reduce waste before looking for new sources – so prior to considering rainwater harvesting methods, you should first reduce your water wastage as much as possible. However, in our rainy country it seems to make sense to collect rainwater for household use before it goes down the drain. This has a threefold advantage. Firstly, it means that less water is taken from reservoirs and rivers. Secondly, it means that rainwater run-off, causing flash flooding, is slightly reduced, and thirdly, it can be collected in situ in a relatively pure state. Although it is possible to filter rainwater to provide drinking water, it is better to use it for toilet flushing, garden watering and other non-potable uses.

The simplest way to make use of rainwater is to put a water butt on all downpipes, and save the water for use in the garden. This is a very cost-effective solution, as the equipment needed will cost only tens of pounds rather than the thousands you would need to spend on a more complicated system.

Water for toilet flushing needs no treatment, though the rainwater store should be covered to prevent the ingress of sunlight and animals. There are now several manufacturers offering packaged rainwater collection, storage and pump units. At present these are quite expensive (approximately £2,000), and you may be better off just carrying out water efficiency measures.

How much water can be collected? An average house in the UK will have around 100m$^3$ of water per year running off its roof (see calculation, right). As the average household use for toilet flushing is 35-70m$^3$ per year (depending on WC flushing efficiency) it is possible, with enough storage, to save this amount of water — worth perhaps £70-£150 per year.

The water from the roof will need to be filtered to prevent dirt, leaves and bird droppings entering and contaminating the flushing system. The best filters (e.g. WISY system at 90% efficiency), operate by surface tension, are self-cleaning, and automatically reject the 'first flush' of contaminated rainwater. Between showers they dry out, so contaminant organisms do not survive. Storage tanks should be sized to contain 1m$^3$ of water per 30m$^2$ of roof. It is preferable to have the storage tank underground, both for aesthetic reasons and as protection against frost. An automatic pumping system is required to deliver water to the WCs. There will normally have to be a mains 'back-up' which can be arranged to come on automatically, and which has to be fitted with approved air gaps to prevent cross-contamination. ●

**A fairly simple calculation can tell you how much rainwater you can harvest, to help you decide if a system (such as below) would be worthwhile: Roof Area (m$^2$) x Annual Rainfall (mm) x System Efficiency (%) x Run-off Coefficient of Roof (a pitched, tiled roof is 0.75) = Annual Collection (l). A tank would normally be sized to store about 5% of this total. For average rainfall figures see www.met-office.gov.uk.**

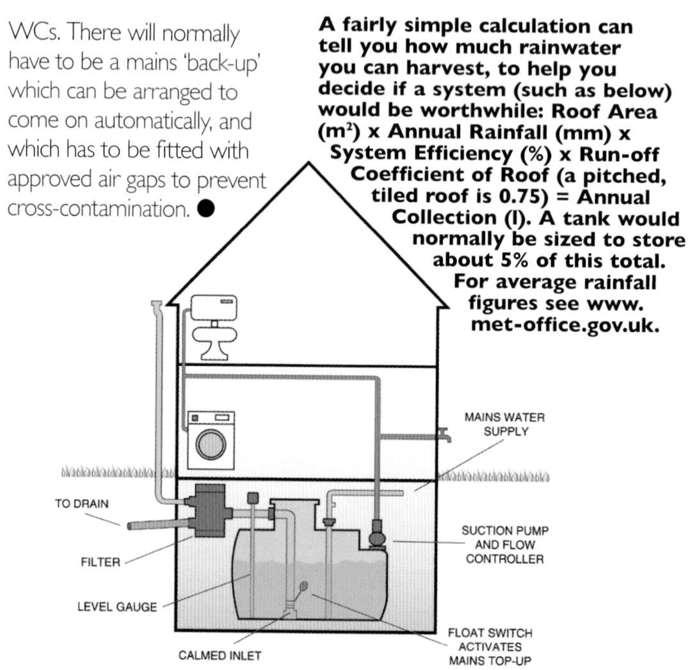

USEFUL CONTACTS: WISY system – Rainharvesting Systems Ltd: 0845 223 5430 www.rainharvesting.co.uk; Freerain: 01636 894906 www.freerain.co.uk; REWATEC: 01844 238111 www.rewatec.co.uk

**Left: A well-designed and positioned conservatory can help warm the whole house for free. Above: Position big windows to the south and small windows on the north side.**

# ENERGY EFFICIENT WINDOWS

Installing energy-efficient windows could save you 25% of your home's heat. So whether you're building or renovating, check out the best options for your project.

Up to 25% of the heat in your house could be escaping through the windows, so it is worth sparing them a thought. But before you do, make sure your walls are well insulated first. If not, then 30% of your heat will be going out that way.

## ORIENTATION

If you have any influence over the position of the windows, make sure the big ones are on the south side and that the north windows are only small. The same applies to conservatories. Well designed and positioned, they will help to warm the house for free!

## FRAMES

It terms of eco glazing, there is only one material for frames: wood. Make sure the timber is from a sustainable source and carries the FSC certificate – temperate (not tropical) hardwood or a durable softwood should be the first choice. Timber frames have had bad press in the past due to poor-quality timber and maintenance, but things have moved on. Even durable softwoods, with a boron pre-treatment requiring no paint, will last 30 years plus.

The ubiquitous PVCu frames are an environmental nightmare. They use a huge amount of energy to produce, the production process releases high levels of dioxins and carcinogenic chemicals and they are non-recyclable.

If you really can't do wood then go for aluminium. They still take a lot of energy to manufacture, but the anodised ones are at least recyclable.

## GLASS

The best, most eco-friendly glazing is salvaged or second-hand double glazing units. They have effectively zero embodied energy and $CO_2$ but are not always practical or possible. The renovator might want to think about secondary double glazing rather than replacing functional single glazed windows. There are lots of options available: from polythene film (from your local DIY store), held in place with double-sided tape and blown with a hair dryer to get the wrinkles out, to metal framed glass, held in place with magnets for easy cleaning. Both will eliminate drafts and significantly improve the U-value.

Double glazed units need to be low-e, soft coated glass, like Pilkington K, with argon gas between the panes. Low-e coating is a thermal reflector that allows heat in but reflects it back into the room. Argon gas improves the performance of the gap and prevents condensation.

Lastly, don't forget the spacer bar. A metal spacer between the panes of glass will act as a thermal bridge, conducting heat away. Ask for double glazed units with insulated spacers.

## CONSERVATION GRADE WINDOWS

Here, the requirements for windows can be tough, but if you have trouble finding good conservation glazing, try contacting English Heritage, Historic Scotland or Cadw (in Wales). ●

## DID YOU KNOW

- **A Scottish building company, the Stewart Milne Group, has developed an affordable zero-carbon green home. Solar panels, compost bins, wind turbines and water efficiency are among the features of the three storey timber-built family house created by PRP Architects for the Aberdeen-based group. The homes are expected to achieve a 5-star rating on the Government's Code for Sustainable Homes.**
- **So many homeowners want renewable energy that the March 2007 grant allocation was taken up in little more than an hour. David Matthews, National Executive Officer of Solar Trades Association, has called on the Chancellor to invest a further £15m to properly fund the LCBP funding program.**
- **Did you know that from April 2008, all building projects valued over £200,000 will be required to have a Site Waste Management Plan? To find out more and to download the guide, visit: www.netregs-swmp.co.uk/simple-guide.pdf**

WORDS: TIM PULLEN

MST
Ashbury
handmade clay roof tiles
....the traditional choice
Designed and produced specifically to give the appearance of an old weathered tile, the ASHBURY range of genuinely hand-made roof tiles look equally at home on renovation or new build projects.
Their manufacture is traditional with each tile being produced entirely by hand. The ASHBURY tile cannot be produced by machinery – the distinctive shades and slight variations of size, camber and texture give the unique appearance so sought after in these days of mass production
Tested and certified by CERAM® Building Technology Ltd to ensure compliance with the relevant British & European Standards, ASHBURY tiles are supplied with an insurance backed 30 Year Guarantee and undergo testing on a regular basis to ensure that the quality standards are maintained.
multi
restoration
cambridge buff
red
huntingdon
Midlands Slate & Tile
Units 9 - 12 Star Industrial Estate, Chadwell St Mary, Essex. RM16 4LR
Tel: 0871 474 3185 Fax: 01375 846 478
email: sales@ashburyrooftiles.co.uk www.ashburyrooftiles.co.uk

# Best selling property titles

Available from all good bookshops, online, or by phone 24 hours on 01480 893833

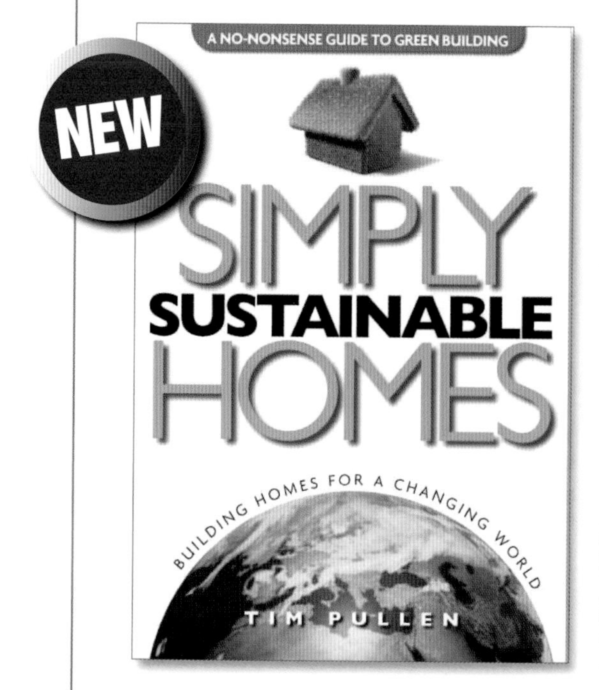


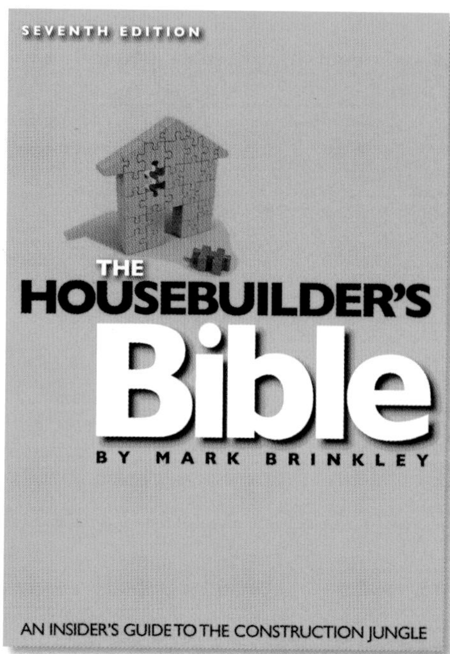

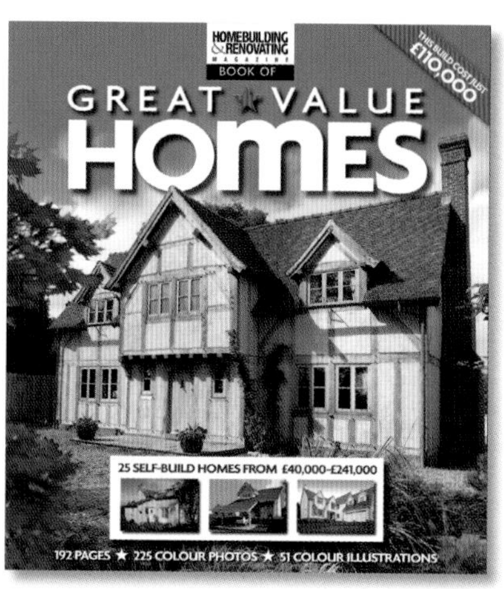

The Granary, Brook Farm, Ellington, Huntingdon, Cambridgeshire PE28 OAE. email: info@ovolobooks.co.uk

# from OVOLO publishing

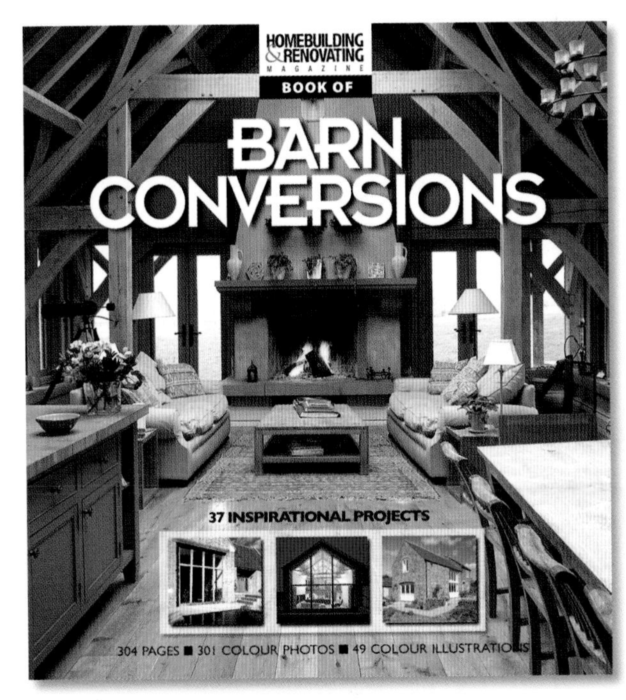


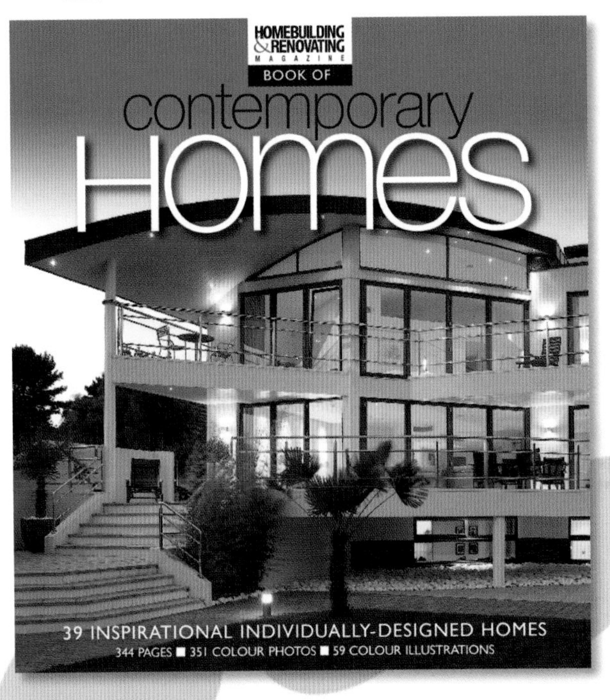

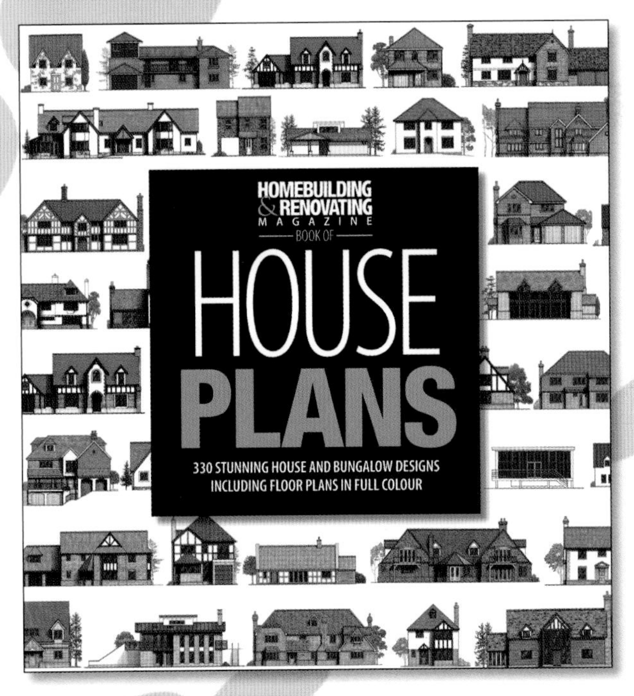